2013 EDITION
MARPOL
HOW TO DO IT

INTERNATIONAL
MARITIME
ORGANIZATION

London, 2013

Published in 2013 by the
INTERNATIONAL MARITIME ORGANIZATION
4 Albert Embankment, London SE1 7SR
www.imo.org

Printed by CPI Group (UK) Ltd, Croydon, CR0 4YY

ISBN 978-92-801-1521-5

IMO PUBLICATION
Sales number: IB636E

Copyright © International Maritime Organization 2013

This publication has been prepared from official documents of IMO, and every effort
has been made to eliminate errors and reproduce the original text(s) faithfully. Readers
should be aware that, in case of inconsistency, the official IMO text will prevail.

H4321

Contents

Part V   Technical aspects of enforcement

Part VI   Organization

Appendices

Figures

Tables

Foreword

MARPOL – How to do it (Manual on the practical implications of ratifying and implementing MARPOL 73/78) was first published by the International Maritime Organization (IMO) in 1993 to provide advice on the process of ratification, implementation and enforcement of the International Convention for the Prevention of Pollution from Ships, 1973, as modified by the Protocol of 1978 relating thereto (MARPOL 73/78). The first revision of the 1993 edition of *MARPOL – How to do it* was published in 2002, followed by a second revision in 2012. The latter was conducted mainly as a result of the total revision of MARPOL Annexes I and II and the entry into force of MARPOL Annexes IV and VI. Annex V was also revised in the meantime.

This publication is the revised and updated version of the 2002 edition of *MARPOL – How to do it*. This edition supersedes the 2002 edition.

Although MARPOL entered into force on 2 October 1983, there are IMO Member States that have not yet ratified MARPOL, or one, or all of its optional Annexes. Of the countries that have ratified MARPOL, there are those that have not managed to implement all of the requirements of the Convention and its related Protocols.

This manual provides useful, practical information to Governments, particularly those of developing countries, on the technical, economic and legal implications of ratifying, implementing and enforcing the MARPOL Convention and its Annexes. The aim is to encourage further ratification, proper implementation and enforcement of the Convention. However, it should be noted that, for legal purposes, the authentic text of the MARPOL Convention and its Annexes, should always be consulted.

The Annexes to MARPOL are living documents that develop through time. It is, therefore, not possible for this manual to reflect texts that are fully up to date and the reader is strongly advised to consult recent updates of the instruments through IMO documents and publications.

1 Introduction to the MARPOL Convention

1.1   The *Torrey Canyon* accident in 1968 prompted a new discussion on ship safety and the protection of the marine environment leading to a decision to develop a comprehensive instrument regarding pollution prevention from ships. The instrument referred to as the International Convention for the Prevention of Pollution from Ships was signed at a diplomatic conference in 1973. The shortened name of that Convention was MARPOL 73. The Convention would enter into force twelve months after the date on which not less than 15 States, the combined merchant fleets of which constitute not less than fifty per cent of the gross tonnage of the world merchant shipping, have become parties to it. After the Amoco Cadiz accident off the coast of Brittany in 1977, it was felt that certain shortcomings in MARPOL 73 should be rectified and a Protocol to the MARPOL 73 Convention was agreed by the International Conference on Tanker Safety and Pollution Prevention (TSPP) in February 1978. At the time of the TSPP conference, the MARPOL 73 Convention had not yet entered into force and could therefore not be amended. To keep it as one Convention it was decided that this Protocol should embrace MARPOL 73.

The Convention, as modified by the Protocol of 1978, was known as MARPOL 73/78.

After the adoption of the 1997 Protocol (see paragraph 1.2 below) it was decided not to add "97" to MARPOL 73/78 but to refer to the Convention just as MARPOL, without any reference to a year. This manual reflects this decision.

1.2   The concern over air pollution was triggered by a growing general awareness that the marine industry should not remain outside the growing worldwide trend to control air pollution sources. This concern resulted in the development of Annex VI, covering a range of air pollutants, which was adopted at a diplomatic conference by means of the 1997 Protocol to the Convention.

1.3   The obligations agreed by the Parties to MARPOL in the articles and regulations relating to different types of ship-generated pollution are contained in the following six Annexes: prevention of oil pollution; pollution from noxious liquid substances carried in bulk; pollution from packaged goods; pollution from sewage; pollution from garbage; and air pollution from ships. These Annexes are explained in more detail in subsequent chapters of this manual.

Status of ratification and implementation of MARPOL

1.4   The Convention entered into force on 2 October 1983, together with the compulsory Annex I (Oil). The, also compulsory Annex II (Noxious Liquid Substances in Bulk) took effect on 6 April 1987. Furthermore, the following optional Annexes have also entered into force: Annex III (Harmful Substances in Packaged Forms) entered into force on 1 July 1992, Annex IV (Sewage) entered into force on 27 September 2003, Annex V (Garbage) on 31 December 1988 and Annex VI (Air Pollution) on 19 May 2005. The status of ratification of MARPOL and its Annexes as of 30 September 2012 is as follows:

Status of MARPOL as of 30 September 2012

MARPOL Annex	Entry into force	Number of ratifications	Fleet (%)
MARPOL Annex I	2 October 1983	152	99.20
MARPOL Annex II	6 April 1987	152	99.20
MARPOL Annex III	1 July 1992	138	97.59
MARPOL Annex IV	27 September 2003	131	89.65
MARPOL Annex V	31 December 1988	144	98.47
MARPOL Annex VI Protocol 1997	19 May 2005	71	94.29

1.5   Although those countries which have ratified the Convention are obliged to implement the requirements of the Convention which include compulsory Annexes I and II and the optional Annexes they have ratified, in respect of their flag ships and all other flag ships in their waters and ports, there still are Parties that are unsuccessful in doing so. The intention of this manual is to assist countries and personnel involved in the planning of the ratification, implementation and enforcement of MARPOL.

1.6   The political desire of a State to accede to or ratify MARPOL is fundamental. Governments may wish to become Parties to MARPOL as a result of:

.1   marine environmental concerns for waters under their jurisdiction;

.2   air quality concerns as they affect the populations or land areas under their jurisdiction;

.3   benefits to their shipowners (worldwide acceptance of ships);

.4   benefits to their ports (means to control pollution); or

.5   concerns for worldwide environment.

1.7   Advice to Governments may come from the public, their own marine or environmental administration bodies or their marine industry.

1.8   It should be recognized that whereas Parties to MARPOL have obligations, they also have privileges. Parties accept the obligation not to discharge residues into the sea or to control the discharges of pollutants to the atmosphere, in return for which they have the privilege of not being polluted by other Parties (if they are polluted, and the pollution occurs within their territorial waters or EEZ, they can prosecute). A non-Party does not accept the obligations to place restrictions upon its ships and, therefore, its ships cannot be prosecuted for failing to comply (except in the territorial waters or EEZ of a Party if apprehended, as the Convention contains the so-called "no more favourable treatment" clause). It has to accept, however, that failure to accept such obligations means that when its own shoreline is polluted or air quality affected it does not have the privilege under MARPOL to insist upon the prosecution of the ship concerned.

1.9   The provisions and the full text of MARPOL and its Annexes are rather complex and it is not easy, without extensive study, to appreciate their full implications directly or to evaluate their impact on the maritime interests or activities of the Administration and the industries of a country.

1.10   The objective of this manual is, therefore, to provide practical information, in an easily understandable form, on the technical, economic and legal implications which may be encountered by Governments and the shipping industry when ratifying and implementing MARPOL. The information provided includes: the obligations that are agreed to when ratifying the Convention, the means of meeting these obligations, the basic marine administration necessary, the legal requirements, the requirements of each Annex of MARPOL, and possible delegation of certain duties by an Administration to other organizations.

1.11   The overall aim is to encourage countries to ratify MARPOL and, in particular, to implement and enforce all the requirements, having first appreciated their obligations, what they need to do, and where concerns may exist.

1.12 The obligations under the Convention and the requirements of the Protocols and Annexes are not reproduced in detail in this manual. An attempt is made to present these obligations and requirements in a straightforward "what needs to be done" manner, along with explanations of the general requirements, making references to MARPOL, IMO resolutions, guidelines and circulars, etc., where the detailed and authoritative texts exist. Furthermore, while the Convention itself has been amended a few times, the Annexes are amended often. In this respect, the latest consolidated edition of MARPOL (2011 edition) provides an excellent reference to the current requirements. In order for a country to consider ratifying or acceding to MARPOL it is essential that its specialists study and analyze this manual. It is also necessary for the maritime authority, shipowners, inspectors and ships' officers and crew to be fully aware of the provisions of MARPOL. This manual sets out these provisions in the following chapters in a practical manner, stating which particular part of the maritime sector needs to take action and on which provision.

1.13 Any reference in this manual to MARPOL means the Convention, its Protocols and its Annexes (see chapter 2).

1.14 Terms used in this manual have, in general, the same meaning as those used in MARPOL (article 2 of the Convention). The following definitions are important for the purpose of this manual and are included here in full Convention wording (article 2). Therefore, unless expressly provided otherwise:

    .1 *Regulation* means the regulations contained in the Annexes to the present Convention.

    .2 *Harmful substance* means any substance which, if introduced into the sea, is liable to create hazards to human health, to harm living resources and marine life, to damage amenities or to interfere with other legitimate uses of the sea, and includes any substance subject to control by the present Convention.

    .3a *Discharge*, in relation to harmful substances or effluents containing such substances, means any release howsoever caused from a ship and includes any escape, disposal, spilling, leaking, pumping, emitting or emptying.

    .3b *Discharge* does not include:

        .1 dumping within the meaning of the Convention on the Prevention of Marine Pollution by Dumping of Wastes and Other Matter, done at London on 13 November 1972; or

        .2 release of harmful substances directly arising from the exploration, exploitation and associated offshore processing of sea-bed mineral resources; or

        .3 release of harmful substances for purposes of legitimate scientific research into pollution abatement or control.

    .4 *Ship* means a vessel of any type whatsoever operating in the marine environment and includes hydrofoil boats, air-cushion vehicles, submersibles, floating craft and fixed or floating platforms.

    .5 *Administration* means the Government of the State under whose authority the ship is operating. With respect to a ship entitled to fly a flag of any State, the Administration is the Government of that State. With respect to fixed or floating platforms engaged in exploration and exploitation of the sea-bed and subsoil thereof adjacent to the coast over which the coastal State exercises sovereign rights for the purposes of exploration and exploitation of their natural resources, the Administration is the Government of the coastal State concerned.

    .6 *Incident* means an event involving the actual or probable discharge into the sea of a harmful substance, or effluents containing such a substance.

Further clarification, or the relevance of these definitions, will be found in the following chapters of this manual.

Part I

Rights and obligations

2 Structure and components of MARPOL

MARPOL is a legal instrument composed of various documents which must be considered as forming a single whole. These documents are described briefly below as they are referred to in this manual and relate to the situation as at July 2012 (full text is reflected in *MARPOL, Consolidated Edition, 2011*).

2.1 International Convention for the Prevention of Pollution from Ships, 1973

Article 1 – General obligations under the Convention

Article 2 – Definitions

Article 3 – Application

Article 4 – Violation

Article 5 – Certificates and special rules on inspection of ships

Article 6 – Detection of violations and enforcement of the Convention

Article 7 – Undue delay to ships

Article 8 – Reports on incidents involving harmful substances

Article 9 – Other treaties and interpretation

Article 10 – Settlement of disputes

Article 11 – Communication of information (article 11(1)(b) was modified by article III of the Protocol of 1978)

Article 12 – Casualties to ships

Article 13 – Signature, ratification, acceptance, approval and accession

Article 14 – Optional Annexes

Article 15 – Entry into force

Article 16 – Amendments

Article 17 – Promotion of technical co-operation

Article 18 – Denunciation

Article 19 – Deposit and registration

Article 20 – Languages

2.2 Protocol of 1978 relating to the International Convention for the Prevention of Pollution from Ships, 1973

Article I   –   General obligations (modifies and adds to the Convention)

Article II   –   Implementation of Annex II of the Convention (delays the implementation of Annex II of the Convention)

Article III   –   Communication of information (modifies article 11(1)(b) of the Convention)

Article IV   –   Signature, ratification, acceptance, approval and accession

Article V   –   Entry into force

Article VI   –   Amendments

Article VII   –   Denunciation

Article VIII   –   Depositary

Article IX   –   Languages

2.3 Protocol I: Provisions concerning Reports on Incidents Involving Harmful Substances

Article I   –   Duty to report

Article II   –   When to make reports

Article III   –   Contents of report

Article IV   –   Supplementary report

Article V   –   Reporting procedures

Resolution A.851(20), as amended by resolution MEPC.138(53): General principles for ship reporting systems and ship reporting requirements, including guidelines for reporting incidents involving dangerous goods, harmful substances and/or marine pollutants. (These guidelines are required to be followed under article V(2) of Protocol I.)

2.4 Protocol II: Arbitration

This Protocol contains articles I to X, which set down the rules for arbitration in accordance with article 10 of the Convention, and relates to settlements of disputes between Parties to the Convention.

2.5 Annex I: Regulations for the prevention of pollution by oil

This Annex contains, at the moment this manual was developed, forty-three regulations. It includes those in the revised Annex which entered into force on 1 January 2007 and all amendments that came or are due to come into force up to and including the amendments adopted during MEPC 63 (March 2012).

The following documents are either an integral part of this Annex or should be considered as such for all practical purposes:

> .1 *Appendices to Annex I*
>
> Appendix I   –   List of oils
>
> Appendix II   –   Form of IOPP Certificate and Supplements
>
> Appendix III –   Form of Oil Record Book
>
> .2 *Unified interpretation of the provisions of Annex I*

.3   *Appendices to the unified interpretation*

Appendix 1 – Guidance to Administrations concerning draughts recommended for segregated ballast tankers below 150 m in length

Appendix 2 – Interim recommendation for a unified interpretation of regulations 18.12 to 18.15 "Protective location of segregated ballast spaces"

Appendix 3 – Connection of small diameter line to the manifold valve

Appendix 4 – Specifications for the design, installation and operation of a part flow system for control of overboard discharges

Appendix 5 – Discharges from fixed or floating platforms

2.6   Annex II: Regulations for the control of pollution by noxious liquid substances in bulk

This Annex contains eighteen regulations; some of these were amended before the Convention came into force, followed by subsequent amendments. A total revised text of Annex II entered into force on 1 January 2007. The following documents are either an integral part of this Annex or should be considered as such for all practical purposes:

.1   *Appendices to Annex II*

Appendix 1 – Guidelines for the categorization of noxious liquid substances

Appendix 2 – Form of Cargo Record Book for ships carrying noxious liquid substances in bulk

Appendix 3 – Form of International Pollution Prevention Certificate for the Carriage of Noxious Liquid Substances in Bulk

Appendix 4 – Standard format for the Procedures and Arrangements Manual

Appendix 5 – Assessment of residue quantities in cargo tanks, pumps and associated piping

Appendix 6 – Prewash procedures

Appendix 7 – Ventilation procedures

.2   *Revised Guidelines for the provisional assessment of liquid substances transported in bulk*

.3   *Provisional categorization of liquid substances, issued on an annual basis on 17 December. This circular is known as the "MEPC.2/Circ"*

2.7   Annex III: Regulations for the prevention of pollution by harmful substances carried by sea in packaged form

This Annex contains eight regulations. The Annex that was adopted with the Convention contained eight regulations but was replaced by a revised version that was adopted in October 1992. A total revised text of Annex III entered into force on 1 January 2010. The appendix, Guidelines for the Identification of Harmful Substances in Packaged Form, is an integral part of the revised Annex, which is implemented through the International Maritime Dangerous Goods Code. By resolution MEPC.193(61), an amended Annex III was adopted, with an entry into force date of 1 January 2014.

2.8   Annex IV: Regulations for the prevention of pollution by sewage from ships

This Annex, at the moment of the update of this manual (second half of 2012), contains thirteen regulations. In 2000 the MEPC adopted a revised Annex IV, however, since Annex IV had not yet entered into force, this adoption took place by agreement by Member States. Finally MEPC 51 adopted a replacement for Annex IV that entered into force on 1 August 2005. At that time, the Annex contained twelve regulations. The thirteenth regulation on port State control on operational requirements, was added later by MEPC 54 and entered into force on 1 August 2007. A further amendment, related to the inclusion of the principle of "special area", was adopted on 15 July 2011 by resolution MEPC.200(62) with an entry into force date of 1 January 2013. Via the same resolution, MEPC.200(62), a consequential regulation 12*bis* on reception facilities for passenger ships in special areas was added to Annex IV. The Appendix, Form of International Sewage Pollution Prevention Certificate, is an integral part of this Annex.

2.9   Annex V: Regulations for the prevention of pollution by garbage

This original Annex contained nine regulations and included several amendments that came into force during its life time. However, a totally revised Annex V, based on the principle of a total prohibition of the discharge of garbage, was adopted by resolution MEPC.201(62) in 2011 with an entry into force date of 1 January 2013. The revised Annex V contains ten regulations and one Appendix.

As a consequence of the revision of Annex V, the Guidelines for the implementation of MARPOL Annex V were also revised and adopted by resolution MEPC.219(63) in 2012. Apart from consequential amendments, it is important to note that the guidelines contain a new chapter on management of cargo residues of solid bulk cargoes.

All references in this manual are to the revised Annex V unless specifically otherwise noted.

2.10   Protocol of 1997, Annex VI: Regulations for the prevention of air pollution from ships

The Protocol of 1997, adding Annex VI to the Convention, came into force on 19 May 2005. While this was a significant addition to the Convention, it was recognized as being only a first step. Consequently, in 2005, work was commenced to strengthen the emission limits and address a number of other identified matters. This work resulted in the development of the revised Annex which was adopted in 2008 by resolution MEPC 176(58) and entered into force on 1 July 2010. All references in this manual are to the revised Annex VI unless specifically otherwise noted. The revised Annex VI contains 18 regulations.

Amendments adopted by resolution MEPC.203(62) in July 2011 added a new chapter 4 to MARPOL Annex VI including five regulations on energy efficiency for ships with an entry into force date of 1 January 2013. The amended Annex VI contains twenty-tree regulations and eight appendices.

The following documents are either an integral part of this Annex or should be considered as such for all practical purposes:

Appendices to Annex VI

Appendix I   –   Form of the International Air Pollution Prevention (IAPP) Certificate

Appendix II   –   Test cycles and weighting factors (relating to the certification of marine diesel engines in respect of nitrogen oxide (NO_x) emissions)

Appendix III   –   Criteria and procedures for the designation of Emission Control Areas

Appendix IV   –   Type approval and operating limits for shipboard incinerators

Appendix V   –   Information to be included in the bunker delivery note

Appendix VI   –   Fuel verification procedure for MARPOL Annex VI fuel oil samples

Appendix VII   –   North American Emission Control Area (Regulation 13.6 and regulation 14.3)

Appendix VIII –   Form of the International Energy Efficiency (IEE) Certificate

NO_x Technical Code 2008 (resolution MEPC.177(58)) Guidelines. There are a number of guidelines referenced in the text of both the revised Annex VI and the NO_x Technical Code 2008.

2.11   Actions required for ratification and implementation

Those concerned with the ratification and implementation of MARPOL should study the documents outlined in this chapter in order to understand the general implications. Further study and in-depth understanding will be necessary for those concerned with particular aspects of ratification and implementation. Guidance on this is given in the subsequent chapters of this manual.

3 Rights and obligations under MARPOL (the Convention and its Protocols)

Many of the articles of MARPOL set down definite requirements. These form the basis towards the regulations of the Annexes and might require specific actions by the Parties. Those articles which do require action are referred to in the following paragraphs in some detail, with the appropriate action indicated. Other articles are mentioned only in order to complete the picture.

3.1   General obligations
(article 1 of the Convention and article I of the 1978 Protocol)

Parties that ratify MARPOL undertake to give effect to its provisions, including those Annexes to which they are bound, these are, the compulsory Annexes I and II and any of the optional Annexes they have accepted, in order to prevent the pollution of the marine environment by the discharge of harmful substances or effluents containing such substances in contravention of the Convention. The means required to meet these obligations are outlined in chapters 4 and 5.

3.2   Definitions
(article 2 of the Convention)

The important definitions contained in article 2 of the Convention have been given in paragraph 1.14 of this manual. Most are straightforward but two definitions are worth mentioning, in order to make it quite clear what MARPOL does and does not cover.

With respect to the definition of *harmful substance*, it should be noted that this definition includes, but is not limited to, substances identified and subject to control by MARPOL and entails those identified in each Annex or provisionally identified through a relevant IMO circular. It does, however, also include other substances that meet the definition of a harmful substance. Means exist within the regulations for dealing with new substances, or with other substances for which the method of carriage by sea has changed.

With respect to *discharge*, it should be noted that substances taken to sea and deliberately dumped under the terms of the Convention on the Prevention of Marine Pollution by Dumping of Wastes and Other Matter, 1972 (the London Convention) or the 1996 Protocol to the London Convention 1972 (the London Protocol) do not come within the scope of MARPOL.

3.3   Application
(article 3 of the Convention)

MARPOL applies to (1) ships entitled to fly the flag of a Party; and (2) ships not entitled to fly the flag of a Party but which operate under the authority of a Party. MARPOL does not apply to any warship, naval auxiliary or other ship owned or operated by a State and used, in the interim, on government non-commercial service. Such ships should, however, be subject to "appropriate measures" and act in a manner consistent with the legislation implementing MARPOL (see chapter 6).

3.4 Violations and sanctions
(article 4 of the Convention)

MARPOL requires Parties to prohibit violations and to provide sanctions under their law and take procedures against offenders. Parties are required to:

.1 apply these to their own-flag ships wherever they may be;

.2 take proceedings against their own-flag ships if sufficient information and evidence of a violation is provided by another Party and inform that Party and IMO of the actions taken;

.3 take proceedings against other ships which commit a violation within their jurisdiction or inform the flag Administration and provide information and evidence of the violation; and

.4 make penalties adequate in severity to discourage violations.

National legislation implementing MARPOL should reflect these requirements, and a marine administration is required to fulfil these obligations.

3.5 Issue and acceptance of certificates
(article 5 of the Convention)

Other Parties should accept a certificate issued under the authority of a Party to MARPOL. A ship that is required to hold a certificate under MARPOL is subject to inspection by officers of a port State which is a Party to MARPOL while in the ports or offshore terminals under its jurisdiction.

This inspection should be limited to verifying that there is a valid certificate on board. If the port State has doubts (clear grounds) that the condition of the ship or its equipment is not in compliance with the certificate, or if it does not carry a valid certificate, the ship should be prevented from sailing until the port State is satisfied that it presents no harm to the marine environment. If action is taken against any ship, its flag State Administration should be informed. Parties to the Convention are required to apply the requirements of MARPOL to non-Party ships to ensure that no more favourable treatment is given to such ships.

Legislation and a marine administration are required to fulfil this obligation.

3.6 Detection of violations and enforcement
(article 6 of the Convention)

Parties to MARPOL agree to co-operate in monitoring compliance with it and detecting violations. Where requested or felt necessary, a coastal or port State shall inspect a ship in order to collect evidence to verify whether it has made a prohibited discharge and, where such a discharge is proved, shall take appropriate measures. A port State shall, in response to a request from another Party, inspect a ship in order to collect evidence or to verify whether it has committed a violation in other waters.

Legislation and a marine administration are required to fulfil this obligation.

3.7 Undue delay to ships
(article 7 of the Convention)

Flag States and port or coastal States should avoid unnecessary or undue delay to a ship on account of measures taken under articles 4, 5 or 6. Where undue delay does occur, the owner or master is entitled to compensation for any loss or damage suffered.

A competent and efficient marine administration is required to fulfil this obligation.

3.8 Reports on incidents involving harmful substances
(article 8 of the Convention)

Parties to MARPOL agree that, without delay, reports of an incident as defined in article 2 including any actual or probable discharge into the sea of a harmful substance or effluents containing such a substance in excess of allowed limits shall be made in accordance with Protocol I. The Parties further agree to make all necessary arrangements for receiving and processing reports, provide details of such arrangements to IMO, report to the Administration of the ship involved and any other State which may be affected, and issue appropriate instructions on reporting to inspection ships and aircraft. This undertaking requires a marine administration and appropriate legislation (see chapter 16).

3.9 Other treaties and interpretation
(article 9 of the Convention)

MARPOL replaces the International Convention for the Prevention of Pollution of the Sea by Oil, 1954 (OILPOL 54). It may, therefore, be necessary for Parties to MARPOL to denounce OILPOL 54 in order to avoid conflict of implementation and to modify any relevant national legislation which they may have in force (see chapter 6). Article 9(3) requires Parties to MARPOL to construe the term *jurisdiction* in the light of international law in force at the time when applying the provisions of MARPOL (see chapter 4).

3.10 Settlement of disputes
(article 10 of the Convention)

Parties to MARPOL agree to arbitration in accordance with Protocol II if necessary in the event of a dispute concerning the interpretation or application of the present Convention (see chapter 17).

3.11 Communication of information
(article 11 of the Convention and article III of the 1978 Protocol)

Parties to MARPOL undertake to provide IMO with documents as follows (for circulation of the information to all Parties):

.1 the texts of laws, orders, decrees and regulations and other instruments which have been published on matters within the scope of MARPOL;

.2 a list of nominated surveyors or recognized organizations which are authorized to act on their behalf in the Administration of matters relating to the design, construction, equipment and operation of ships carrying harmful substances in accordance with the provisions of the regulations, for circulation to the Parties for the information of their officers. The Administration shall therefore notify IMO of the specific responsibilities and conditions of the authority that is delegated to nominated surveyors or recognized organizations (this information was amended by article III of the 1978 Protocol);

.3 a sufficient number of specimens of their certificates issued under the provisions of the regulations;

.4 a list of reception facilities, including their location, capacity and available facilities and other characteristics;

.5 official reports or summaries of official reports in so far as they show the results of the application of MARPOL;

.6 an annual statistical report, in a form standardized by IMO, of penalties that have actually been imposed for infringement of MARPOL; and

.7 regarding communication to IMO, attention should be given to the Global Integrated Shipping Information System (GISIS) developed in order to promote public access to sets of data collected by the IMO Secretariat. The management of the rights to access and use of the GISIS electronic reporting facilities is left to the discretion of Member States. For detailed information, reference is made to Circular letter No.2639 of 8 July 2005.

This undertaking requires a marine administration capable of producing the required documents.

3.12 Casualties to ships
(article 12 of the Convention)

Parties to MARPOL undertake to investigate any casualty occurring to any of their ships which are subject to the regulations, if such casualty has produced a major harmful effect on the marine environment. Parties also undertake to provide IMO with the findings of such investigations if such information may assist in determining what changes to MARPOL might be desirable. In this respect, reference is made to the GISIS database referred to in paragraph 3.11.7 above.

This undertaking requires that there is a marine administration that is capable of conducting investigations into casualties and providing appropriate reports.

3.13 Signature, ratification, acceptance, approval and accession
(article 13 of the Convention and article IV of the 1978 Protocol)

MARPOL remains open for accession.

Although the articles permit States to become Parties by "signature", this facility was to enable Governments of States to declare their intent to work in good faith to ratify MARPOL. Now that MARPOL is in force, it is no longer open for signature. The only means of becoming a Party now is by accession. Accession requires action by the Government of a State, which is further explained in chapter 6.

3.14 Optional Annexes
(article 14 of the Convention)

A State may, at the time of signing, ratifying, accepting, approving or acceding to MARPOL, declare that it does not accept any one or all of Annexes III, IV, V and VI (optional Annexes) but may accept any such optional Annex later. For Annex VI, acceptance is achieved via acceding to the 1997 Protocol.

A State, which has not accepted an optional Annex, shall be under no obligation in respect of matters related to that Annex. Parties to MARPOL shall otherwise be bound by any Annex in its entirety.

3.15 Entry into force
(article 15 of the Convention and article V of the 1978 Protocol)

These articles provide the conditions and timing of entry into force of MARPOL, its optional Annexes and its amendments.

3.16 Amendments
(article 16 of the Convention and article VI of the 1978 Protocol)

This article provides the procedures for amendments to MARPOL. It should be noted that only Parties to MARPOL may propose, make decisions on and vote on amendments. The same procedure applies to the amendments to the so-called optional Annexes.

3.17 Promotion of technical co-operation
(article 17 of the Convention)

Parties to MARPOL undertake to provide, in consultation with IMO and other international bodies, support for those Parties which request technical assistance on matters relating to MARPOL such as training of technical and scientific personnel, supply of equipment for monitoring and control, reception facilities, adoption of prevention and control measures, and the encouragement of research. It is important to note that a request for technical co-operation can only be made by a Party to the Convention (see chapter 24).

3.18   Denunciation
(article 18 of the Convention and article VII of the 1978 Protocol)

Parties to MARPOL may terminate their obligations by notifying the Secretary-General of IMO.

3.19   Deposit and registration
(article 19 of the Convention and article VIII of the 1978 Protocol)

The 1973 Convention and the 1978 Protocol are deposited with the Secretary-General of IMO, who will inform all existing Parties of new Parties or denunciations (see chapter 6).

3.20   Languages
(article 20 of the Convention and article IX of the 1978 Protocol)

Authentic texts are established in the English, French, Russian and Spanish languages with official translations in the Arabic, German, Italian and Japanese languages. For the 1997 Protocol, authentic texts are established in Arabic, Chinese, English, French, Russian and Spanish languages.

3.21   Protocol of 1978

The purpose of this Protocol was to enable Annex I of the Convention (Regulations for the Prevention of Pollution by Oil) to be amended and implemented as soon as possible and to permit delay in implementing Annex II (Regulations for the Control of Pollution by Noxious Liquid Substances in Bulk) because of technical difficulties. In the event, Annex I came into force in 1983, some three and a half years ahead of Annex II. All articles of the Protocol have been mentioned in the foregoing sections of this chapter and will not be explained further.

3.22   Protocol of 1997

Whereas Annexes I – V are part of the 1973 Convention either as obligatory Annex to the Convention (Annexes I and II) or as optional Annex (Annexes III, IV and V) as originally adopted, Annex VI is an entirely new Annex and was therefore introduced by means of the 1997 Protocol, article 2. Article 3 links the 1997 Protocol, and Annex VI, to the preceding 1973 Convention. Article 6 provides both the specific entry into force conditions of the Protocol and for those States which agree to be bound by its provisions after that date – three months after the date of deposit of the instrument of accession. The other articles have the same meaning as the corresponding articles previously mentioned.

3.23   Reception facilities

The provision and use of reception facilities for ship-generated wastes and, where applicable, cargo residues are integral parts of the rights and obligations of MARPOL. Experience has shown that both provision and adequate use of reception facilities are complex issues. Chapter 15 of this manual deals with the provision of reception facilities in more detail, as do the provisions in Annexes I, II, IV, V and VI (see also paragraph 25.8 of this manual).

3.24   MARPOL regulations on enforcement

The obligations discussed in the preceding sections set the framework for mutuality and co-operation in the enforcement of MARPOL. In order for ships to be in compliance with the basic framework for the control of operational pollution, a Party must also ensure that certain requirements are met in respect of ship construction, equipment, documentation and operational procedures. These requirements are set out in the technical Annexes. The enforcement rights and obligations of a Party under the technical Annexes of the Convention are detailed in the ensuing chapters.

MARPOL establishes the rights and obligations of enforcement applicable to any State Party to the Convention as a flag, port or coastal State, subject to relevant safeguards. The distinctions of jurisdiction and enforcement for the flag State, the port State and the coastal State are further developed in chapter 4 of this manual.

3.25 Actions necessary for implementation and enforcement

In order to enforce the provisions of MARPOL, a State Party must give full effect to the provisions of the Convention under national law. This includes the passing of enabling regulations in respect of all the technical Annexes to which the State is bound, and the incorporation of a framework of sanctions against violations within the jurisdiction of a State Party. A summary of the key actions which any State Party to MARPOL may undertake in order to comply with the requirements of the Convention is presented in figure 1.

1	Accede to MARPOL
2	Give effect to Annexes I and II
3	Give effect to the other Annexes accepted or given force by national law
4	Prohibit violations
5	Provide sanctions
6	Take proceedings
7	Inform Parties concerned
8	Inform IMO
9	Inspect ships
10	Monitor compliance
11	Avoid undue delay to ships
12	Report on incidents
13	Provide IMO with documents (article 11 of the Convention)
14	Investigate casualties involving pollution and report findings
15	Ensure provision of adequate reception facilities

Figure 1 – *Actions necessary in implementing and enforcing MARPOL*

3.26 Special areas (SAs)

Due to specific oceanographic, ecological and shipping characteristics of some sea areas, the Convention has established "special areas". The discharge of oil or oily mixture, noxious liquid substances carried in bulk, sewage and garbage (Annexes I, II, IV and V of the Convention) is subject to control in special areas. Precise details are contained in the Convention. In designing enforcement strategies for the Convention, States should take into consideration the relative importance of compliance in these special areas. The establishment of an area as a special area will only take effect upon sufficient receipt of notifications of the existence of adequate reception facilities by IMO, from Parties whose coastlines border the relevant special area. For the latest status of the special areas, it is advised to consult the IMO website (www.imo.org).

The special areas established under the Convention are as follows

- **Annex I:** the Mediterranean Sea area, the Black Sea area, the Baltic Sea area, the Red Sea area, the "Gulfs" area, the Gulf of Aden area, the Antarctic, the North West European Waters, the Oman area of the Arabian Sea and the Southern South African waters.

- **Annex II:** the Antarctic area.

- **Annex IV:** the Baltic Sea area.

- **Annex V:** the Mediterranean Sea area, the Black Sea area, the Baltic Sea area, the Red Sea area, the "Gulfs" area, the North Sea area, the Antarctic area and the Wider Caribbean region.

3.27 Emission control areas (ECAs)

In a similar manner to the "special areas" mentioned above, in accordance with Annex VI, appendix III, certain areas have been designated "emission control areas" (ECAs). These are areas where it has been demonstrated that the emissions to atmosphere from international shipping have a particularly adverse effect in adjacent land areas on either public health or the wider environment. In an ECA, lower limits are applied to nitrogen oxide (NO_x) emissions or sulphur oxide (SO_x) and particulate matter emissions or all three types of emissions. Further details regarding these ECAs, including aspects relating to their entry into effect dates, are provided in chapter 14 of this manual.

At the time of development of this manual, the established ECAs are:

For NO_x control (Tier III – generally applicable to ships constructed on or after 1 January 2016 operating in a designated ECA):

    .1   North American area – this covers up to 200 miles from the coastlines of much of the USA (including Hawaii) and Canada together with the territorial waters of Saint-Pierre-et-Miquelon;

    .2   United States Caribbean Sea area.

For SO_x and particulate matter control:

    .1   Baltic Sea area;

    .2   North Sea area;

    .3   North American area – as above (with an entry into force date of 1 August 2011);

    .4   United States Caribbean Sea area (with an entry into force date of 1 January 2013).

Additional emission control areas for NO_x (NECA) or for SO_x (SECA) or for both NO_x and SO_x (ECA) may be established over time and hence the relevant IMO documentation should be referred to for the current status and extent of these areas.

4 Jurisdiction

4.1 Overview

4.1.1 MARPOL and the International Law of the Sea

MARPOL, article 9(3) requires that jurisdiction be construed in light of international law in force at the time of application or interpretation of MARPOL. Such international law, as set forth in the 1982 United Nations Convention on the Law of the Sea (UNCLOS), describes the circumstances, safeguards, and geographical zones of coastal, flag and port State jurisdiction, among other things. Thus, for many Parties to MARPOL, international law affects how MARPOL will be enforced. For ease of reference, MARPOL provisions which are complementary to, or require interpretation in light of, the provisions of UNCLOS are cross-referenced in the overview below:

MARPOL/UNCLOS cross reference

MARPOL section	UNCLOS section
1(1)	94, 217(1)
4(2)	21(1), 56(1)(B), 211, 220, 228, 231
4(3)	217(7)
5	217(3)
5(2)	217(2)
6	218
7	226(1), 232
9(3)	91, 217, 220, 218
10	287

4.1.2 Forms of jurisdiction

In discussing jurisdiction it is essential to distinguish between a State's competence to prescribe legislation for individual ships (legislative jurisdiction), and its competence to enforce legislation thus prescribed (enforcement jurisdiction). Secondly, the legislative or enforcement jurisdiction that a State has in respect of a particular ship varies depending on whether it is a flag, coastal or port State.

The legislative and enforcement jurisdiction and responsibilities of flag, coastal and port States are more fully described in the sections below. However, all States, in implementing MARPOL, are required to apply its requirements so that ships of non-Parties to MARPOL receive no more favourable treatment than ships of Parties. See MARPOL, article 5(4).

4.2 Flag State jurisdiction

4.2.1 Legislative jurisdiction and obligations of the flag State

The primary responsibility for the implementation of international standards and regulations relating to ships is held under the jurisdiction of the flag State. These international standards include MARPOL.

MARPOL, article 1(1) requires Parties to give effect to the Convention including all Annexes by which they are bound. In taking these measures, flag States are required to conform to generally accepted international regulations, procedures and practices. For MARPOL, these regulations, procedures and practices can be found in the articles, Protocols, Annexes, unified interpretations of the Annexes, and IMO circulars and publications relating to MARPOL.

The flag State has two main responsibilities in ensuring that its ships comply with the technical standards set by MARPOL. First, it must survey and inspect ships at periodic intervals. Second, it must issue relevant certificates showing that, according to their size and type, its ships are in compliance with the Annexes they are Party to.

Administrations are required to conduct surveys of certain ships to determine the existence of certain equipment and procedures mandated by MARPOL, to approve the equipment and procedures, and to ensure that the condition of the ship as stated on any certificate issued conforms with the actual condition of the ship. The survey requirement under flag State control is the most detailed. Surveys are required under Annexes I, II, IV and VI of the Convention. Ships and equipment are subject to mandatory initial, annual, intermediate, renewal and periodical surveys (see MARPOL, Annex I, regulation 6; Annex II, regulation 8; Annex IV, regulation 4 and Annex VI, regulation 5). Unscheduled inspections for compliance with Annex I are also authorized, as are surveys after an accident occurs or a defect is discovered that could affect compliance with Annexes I, II, IV or VI. It should be noted that under Annex IV no annual and no intermediate surveys are required.

After a survey, Administrations are required to ensure that ships under whose authority they are operating, carry and are issued certificates demonstrating compliance with various MARPOL, Annexes (see Annex I, regulations 7–10; Annex II, regulations 9-10; Annex IV, regulations 5–8 and Annex VI, regulations 6–9). Where a flag State authority finds non-compliance with Annex I, II, IV or VI, the authority withholds issuing the applicable certificates or shall withdraw any certificates previously issued until compliance with the terms of the certificate (and thus the substantive terms of the Convention) is achieved. See MARPOL, Annex I, regulation 6.3.4, Annex II, regulation 8.2.6, Annex IV, regulation 4.6 and Annex VI, regulation 5.3.3. Further, flag States must prohibit non-compliant ships from sailing until they can proceed to sea in compliance with MARPOL, or can proceed to the nearest repair yard without presenting an unreasonable threat to harm the marine environment (see MARPOL, article 5(2), Annex I, regulation 6.4.1, Annex II, regulation 8.3.1, Annex IV, regulation 4.7 and Annex VI, regulation 5.3.4). Certificates provide prima-facie evidence that the ship complies with the requirements of MARPOL: each "shall be accepted by other Parties and regarded for all purposes covered by the present Convention as having the same validity as a certificate issued by them". See MARPOL, article 5(1).

4.2.2   Enforcement jurisdiction of the flag State

MARPOL obliges Administrations to prohibit violations of the Convention. Further, Administrations are required to investigate reports of violations, and if sufficient evidence is available, institute proceedings in accordance with its law (see MARPOL, articles 4(1) and 6(4)). If the alleged violation occurs in the jurisdiction of another coastal or port State, the flag State shall receive evidence from the affected Party, and may ask for further or better evidence to enable proceedings to be brought.

Administrations are further required by MARPOL, article 4(1) to establish appropriate sanctions for violations.

In some instances, allegations of a violation are reported by a State other than the flag State. After receiving these reports, the flag State must inform the reporting State and IMO of the action taken by it in response to the allegation. See MARPOL, article 4(3).

4.3   Jurisdiction of the coastal State

4.3.1   Legislative jurisdiction of the coastal State

MARPOL, article 4(2) requires that coastal States (Parties) prohibit violations and establish sanctions for any violation of the Convention that occurs in their jurisdiction. In the territorial sea, the coastal State enjoys sovereignty, and with it the power to apply national law, subject to conformity with generally accepted principles of international law.

Some coastal States claim an Exclusive Economic Zone (EEZ) in accordance with the principles described in UNCLOS Part V. In the EEZ, the coastal State has jurisdiction with regards to the protection and preservation of the marine environment. Thus, MARPOL may be applied consistent with this grant of jurisdiction.

4.3.2 Enforcement jurisdiction of the coastal State

MARPOL does not specifically address the enforcement jurisdiction of coastal States, other than the requirement contained in article 4(2) either to cause proceedings to be taken in accordance with coastal State law, or to furnish information and evidence regarding the violation to the flag State Administration of the ship involved.

If a foreign ship is voluntarily in a port of an affected coastal State, that State may institute proceedings for any violation of MARPOL that has occurred within the territorial sea or EEZ of the coastal State.

If a foreign ship is navigating in the territorial sea of a coastal State and there are clear grounds for believing that, during its passage in the territorial sea, the ship has committed a violation of the coastal State's laws implementing MARPOL, the coastal State may inspect the ship, and if the evidence so warrants, institute proceedings.

If a foreign ship is navigating in the territorial sea or EEZ of a coastal State and there are clear grounds for believing the ship has committed a violation of the coastal State's laws implementing MARPOL in the EEZ, the coastal State may require the ship to give *its identity, ports of call and other relevant information* to establish if a violation has occurred. If that violation results in a substantial discharge or threatens significant pollution to the marine environment in the EEZ, the coastal State may inspect the ship if the ship refuses to give the information requested, or the information requested is manifestly at variance with the evidence. When there is clear objective evidence that the violation has, in fact, resulted in a substantial discharge or threatens significant pollution to the marine environment in the EEZ, the coastal State may institute proceedings. However, for violations occurring in the EEZ, the coastal State must suspend proceedings and transfer them to the flag State if the flag State also institutes proceedings within specified time limits.

Coastal States instituting proceedings against a foreign ship must notify the flag State and IMO of any measures or proceedings taken for violations of MARPOL, except that for violations occurring in the territorial sea, only proceedings need be reported.

4.4 Jurisdiction of the port State

4.4.1 Legislative jurisdiction of the port State

MARPOL, article 4(2) requires Parties to prohibit violations of the Convention and provide for sanctions for violations within their jurisdiction, including in ports. If conditions for entry into port involve requirements for the prevention, reduction and control of pollution, however, the port State must give due publicity to such requirements to IMO.

4.4.2 Enforcement jurisdiction of the port State

MARPOL does not leave the question of compliance enforcement to the flag State alone. A port State can exercise enforcement jurisdiction against any ship visiting its ports to ensure compliance with, and to detect violations of MARPOL. The jurisdictional rights of port States include:

4.4.2.1 *Inspection of certificates*

Pursuant to MARPOL, article 5, ships required to hold certificates issued pursuant to the Annexes to MARPOL are subject to having those certificates ready for inspection by authorized port State control officers (PSCO). This certificate inspection is initially limited to verifying that a valid certificate is on board the ship. However, if there are clear grounds for believing the condition of the ship or its equipment is not substantially in compliance with the terms of the certificate, a more thorough inspection may be conducted.

Where non-compliance is revealed, port States must not allow such ships to sail unless they can do so without presenting an unreasonable threat of harm to the marine environment. Frequently, the port State's most effective sanction is therefore to detain the ship in port until it can be repaired to a suitable standard or directed to a repair yard. The flag State shall be informed immediately of the steps taken. Further, the ship may be reported to the flag State for appropriate action, or the port State may initiate proceedings under its own law for any violation which arises from non-compliance with the Convention (see MARPOL, article 4(2). However, the port State must not unduly delay ships (see MARPOL, article 7).

4.4.2.2 *Inspection to detect violations of the discharge standards*

Pursuant to MARPOL, article 6, a ship in any port may be subject to inspection for the purpose of verifying whether the ship has discharged any harmful substances in violation of the MARPOL regulations. A report must be made to the flag States when a discharge violation is indicated and flag States must then bring proceedings if satisfied that the evidence is sufficient. The requirement for flag States to bring proceedings, however, does not pre-empt the right of port States to bring their own proceedings for violations which occur in the territorial sea or internal waters of the port State, or which cause major damage to the port State. See MARPOL, article 4(2).

If a port State receives a request from any Party, it may inspect a ship to investigate whether the ship has discharged harmful substances or effluents containing harmful substances in excess of the quantity allowed in any place, provided the port State also receives sufficient evidence from the requesting Party that the discharge occurred. See MARPOL, article 6(5).

4.4.3   Extended port State enforcement jurisdiction

In addition to the power to investigate and initiate proceedings for violations occurring within the port State the port State has the power to investigate and take proceedings for discharge violations wherever they have taken place. This power covers violations within the internal waters, territorial seas and exclusive economic zone (EEZ) of another State. In these instances, the port State may only take proceedings in response to a request from States damaged or threatened by the discharge violation, or in response to a request by the flag State concerned. However, the affected coastal State does enjoy a right of pre-emption, that is, it can request that any investigation, records or proceedings by the port State be suspended and transferred to the coastal State.

Except in cases of violations occurring inside the territorial sea of a coastal State, or a violation causing major damage to the coastal State, the flag State can insist on taking control of any proceedings for any violation which occurs outside the territorial sea of the coastal or port State. The flag State must continue the proceedings, and it loses the right of pre-emption if it repeatedly disregards its obligation of effective enforcement of international regulations.

4.5   No oppressive exercise of authority

In exercising any of these enforcement powers, coastal or port States must observe certain safeguards whose purpose is to prevent oppressive exercise of their authority. In particular, they must not act in a discriminatory fashion. These safeguards include:

4.5.1   Prompt investigations

States shall not delay a ship longer than is essential for purposes of investigation. Further, if an investigation or inspection reveals a violation, a ship must be released once a reasonable bond or other financial security is posted, unless releasing the ship would pose an unreasonable threat of damage to the environment. Failure to release a ship once it has posted a bond or acceptable financial security (assuming no threat of environmental damage) may result in implementation of dispute resolution proceedings held in accordance with MARPOL, article 10. States that unduly delay or detain ships shall be liable for compensating for loss or damage suffered. See MARPOL, article 7(2).

4.5.2 Sanctions

In imposing sanctions for violations, States may impose only monetary penalties, except in the case of a wilful and serious discharge in violation of MARPOL that occurs in the territorial sea of the affected State.

States Parties to MARPOL should take due care to reflect the above-mentioned provisions in enabling national legislation.

4.6 Enforcement regime

MARPOL obligates Contracting Parties to enforce laws and regulations relating to the prevention, reduction and control of pollution of the marine environment from ships flying their flag and from foreign ships operating within their jurisdiction. Ensuring compliance takes many forms, the ultimate of which is prosecution. While States Parties are required to provide adequately for the prosecution of MARPOL violations, there does not appear to be any consensus internationally as to the characterization of MARPOL offences. Civil law jurisdictions usually treat such offences as "minor", "serious" or "aggravated", that is, according to the degree of the severity of the damage caused. Some common-law jurisdictions treat these as "strict" or "absolute" liability offences. Some jurisdictions adopt the so-called "half-way house" approach for MARPOL offences. In this characterization, the offence would be treated as one of "strict liability", subject to affording the violator the defence of due diligence, i.e., the onus would be on him to show that he took all reasonable precautions and exercised all due diligence to avoid the commission of the offence.

4.7 Prosecuting offences

Prosecution strategy and tactics will depend largely on the applicable legal system under which a violation is prosecuted, the underlying national law implementing MARPOL, the provisions violated, and the nature of the violation. Nevertheless, there are some common concerns that should be considered regardless of these variables. These include jurisdiction, presentation of evidence, standards of proof of violation, the nature of sanctions imposed (administrative, civil and penal), and the rights of the accused. A Government seeking to proceed against a person accused of violating MARPOL must first establish jurisdiction over the accused or any property that may be subject to seizure and forfeiture.

Governments should be mindful of their own judicial, administrative and enforcement constraints when developing systems for prosecuting MARPOL offences. While severe sanctions can be effective deterrents, they can also require an expensive and burdensome proof process which is, in itself, a deterrence to the policing authorities. Governments should consider providing a range of options to the enforcement authorities for the prosecution of violations. This range could include warnings on the spot for minor violations; "tickets" on the spot for minor violations which would justify more than a warning; and citations to be handled at three levels: at magistrate level, higher civil court level, and at criminal court level.

Certain common law jurisdictions use administrative sanctions in lieu of criminal sanctions for environmentally related and regulatory offences such as are applicable under MARPOL. Two key factors have promoted this development. First, the cases can be disposed of quickly rather than going through a time-consuming court process and are also less costly. Second, the use of administrative sanctions obviates the need for the process of complex methods of proof that is required to establish guilt for criminal offences. Under a system of administrative sanctions, a Government may wish to provide the power to compound offences or provide for the increase of the stated fine, under certain instances.

Sanctions, be they administrative or penal in nature, would, by and large, consist of fines. It would be reasonable to provide for a range with a minimum and maximum level, with the exact amount of the fine being dependent on the severity of the offence.

4.8   The Administration

The primary responsibility for ensuring compliance with MARPOL rests with the Administration as defined in article 2(5). The Administration must have the necessary authority and resources to administer, enforce and ensure compliance with the provisions of the Convention. The Administration may, however, under the restraints imposed by the Convention, entrust certain tasks to surveyors nominated for the purpose or to organizations recognized by it. Nevertheless, the Administration shall fully guarantee the completeness and efficiency of the tasks so delegated and shall undertake to ensure the necessary arrangements to satisfy this obligation.

Part II

Preparations

5 Means of meeting obligations

5.1 Participation

Accession to MARPOL and its implementation require the participation of the following:

.1 Government of the State (the political body having power to conclude international agreements)

.2 Administration – legal

.3 Administration – maritime

.4 Shipowners

.5 Port authorities

Each sector should know exactly what its institutional rights, obligation and responsibilities are, and the responsibilities of its staff and the requirements to be imposed on ships and ports.

5.2 Government of the State

The political desire of a State to accede to or ratify MARPOL is fundamental. Governments may wish to become Parties to MARPOL as a result of:

.1 marine environmental concerns for waters under their jurisdiction;

.2 concerns over air quality, which affects the populations, or land areas under their jurisdiction;

.3 benefits to their shipowners (worldwide acceptance of ships);

.4 benefits to their ports (means of control of pollution); or

.5 concern for the worldwide environment.

Advice to Governments may come from the public at large, from their own maritime or environmental administration bodies and from their maritime industry.

It should be recognized that whereas Parties to MARPOL have obligations, they also have privileges. Parties accept the obligation not to discharge harmful substances into the sea or to control the discharge of pollutants to the atmosphere, in return for which they have the privilege of not being polluted by other Parties. (If they are, and the pollution occurs within their territorial waters, they can prosecute.) A non-Party does not accept the obligations to place restrictions upon its ships and, therefore, its ships cannot be prosecuted for failing to comply (except in the territorial waters of a Party if apprehended and Parties shall apply the requirements of the present Convention as may be necessary to ensure that no more favourable treatment is given to such ships). It has to accept, however, that failure to accept such obligations means that when its own shoreline is polluted or air quality affected it does not have the privilege under MARPOL to insist upon the prosecution of the ship concerned.

5.3   Administration – legal

Once the political desire has been established and a decision made to become a Party, it is necessary to consider the means of ratifying and implementing MARPOL. This aspect is covered in chapter 22.

5.4   Administration – maritime

The marine administration will have by far the greatest administrative task in the implementation of MARPOL. It is likely that this body will provide advice to the legal branch and the Government of a State on one hand, and will advise the shipping industry and port authorities on the other. These aspects are covered in chapter 22.

5.5   Shipowners

Shipowners will need to design, construct and equip ships and train seafarers, especially their merchant marine officers, in order to meet the requirements of MARPOL. An outline of these requirements is given in part IV of this manual, implementing the regulations, chapters 9 to 14, under the heading of the respective Annexes of MARPOL.

5.6   Port authorities

The main concern of port authorities, next to enforcement in port, will be the provision of adequate reception facilities. The requirements are given under the heading of the respective Annexes of MARPOL in chapters 9 to 14 and chapter 15 of this manual.

5.7   Consultation

When a State is considering acceding to or ratifying MARPOL it is likely that many organizations that fall within the foregoing categories (paragraphs 5.2 to 5.6) will need to be consulted in order to be properly prepared to implement and enforce all of the obligations and requirements.

5.8   Impact of MARPOL

When considering the necessary means of meeting its obligations under MARPOL, a State should recognize that the impact will vary according to whether it is a flag State, a port State or a coastal State. Most States will be all three, but some may be very large flag States but have little in the way of port or coastal State responsibilities. The impact of MARPOL will also vary with the trade of a port State and the type of ship for a flag State. For example, a flag State may have a large fleet of chemical tankers and consequential responsibilities under MARPOL Annex II but little concern as a port State because it does not import or export noxious liquid substances in bulk (see further comments under chapters 9 to 14 of this manual).

5.9   Obligations

All sectors involved with MARPOL need to consider and meet their obligations with respect to:
- preparation of legislation, including regulations
- survey
- inspection
- design and constructional requirements
- equipment requirements
- operational requirements
- documentation
- procedures
- agreements with other Governments.

5.10   Developing a compliance strategy for the Convention

5.10.1 Why compliance?

Article 1(1) of MARPOL requires all Parties to the Convention to "undertake to give effect to the provisions of the present Convention and those Annexes thereto by which they are bound, in order to prevent the pollution of the marine environment by the discharge of harmful substances or effluents containing such substances in contravention of the Convention". In accordance with this obligation, a Party to MARPOL will need to implement a range of monitoring, compliance and enforcement mechanisms to give force and effect to the Convention. Compliance with the Convention should primarily focus on preventing pollution, and not simply on apprehending and punishing violators. The extent to which education, incentives, monitoring and policing programmes are used by a State to ensure compliance with MARPOL depends upon the type of jurisdiction that the State enjoys over a ship (see chapter 4).

5.10.2 Strategies for ensuring compliance

An effective compliance programme should incorporate all of the following elements:

.1   compliance monitoring through routine inspections, surveys, and/or examinations (see chapter 21);

.2   detection and policing patrols (see chapters 20 and 21);

.3   reporting procedures and incentives, including incentives for self-reporting (see paragraph 20.6);

.4   adequate investigations of violations reported or otherwise detected (see chapter 20);

.5   a system of adequate sanctions in respect of violations (see chapter 7);

.6   education and public awareness programmes (see paragraph 5.10.3); and

.7   co-operation and co-ordination with other Parties (see paragraphs 3.6 and 5.10.4).

A compliance programme should be adaptable enough to allow compliance priorities to respond to prevailing circumstances. One or more of its elements may be more salient for a State Party depending on key variables, including the state of the national fleet, the type of ships calling at ports of the State Party, the emergence of new equipment/procedural Convention standards, the availability of human and technological resources within the marine administration, and the familiarity of relevant stakeholders with the Convention.

In setting priorities for a compliance strategy, the marine administration would need to have an idea of which ships have the highest potential for being in violation, or where a violation would be most significant (see also chapter 21).

5.10.3 Public participation

Any compliance strategy should take into consideration that resources spent on education and prevention will save resources that might have been spent on prosecution. Education and prevention strategies are necessary to sensitize all potential "observers" and "witnesses" about how they can assist in protecting the marine environment. In this way, they may prove a cost-effective resource for States Parties with limited financial or policing resources. Public awareness and participation can also greatly facilitate the reporting of violations and may also prevent violations that would never have been detected or prosecuted. One means of encouraging reporting of potential MARPOL violations by the general public has proven successful, particularly for incidents involving oil or garbage that are easily observable in the water: bounties. It should be noted that some States' domestic laws do not permit bounties. For other States, bounties are inconsistent with their enforcement objectives. However, for those that wish to consider such incentives, States which utilize bounties have found them to be useful in the detection and prosecution of MARPOL violations.

Public awareness is likely to be most effective when the programmes are directed at site-specific targets, for example, marinas and fishing docks, and at their respective Administrations and users. Public education can also engender more positive behaviour from consumer-oriented operations such as the cruise lines. There should be specific information provided that pollution of the marine environment by oil, hazardous materials, sewage, garbage and air pollution is against the law. There should also be cautionary training that some pollution from these sources may be hazardous and that only officials should come into contact with the material to verify its nature. Because of the potential distribution of, and sensitivity to, air pollution communities and areas concerned may be remote from port activity or the coast, and therefore the spread of the necessary information needs to be that much wider. Included should be information as to how air pollution from shipping contributes to the overall air pollution encountered. Most importantly, the public should know how to contact the official authority responsible for initiating the appropriate response.

5.10.4 Co-operation and co-ordination of port State control

Articles 6 and 8 of MARPOL, as well as several important IMO resolutions, lay the ground work for the doctrine of co-operation and interchange as a mutual effort of enforcement among Parties to the Convention.

Such co-operation can be an effective tool in fostering clarity and harmony in implementation and compliance objectives, in collecting evidence, and in enforcement procedures. Co-operation may take several forms, such as joint investigations of violations, supplying information about a particular ship, gathering evidence of a violation, and prosecuting flag State ships within the jurisdiction of another country for provable MARPOL violations. Reciprocal arrangements in respect of investigations and compliance monitoring will be particularly valuable for Parties which are geographically proximate and/or which share common mechanisms for enforcement. Such arrangements can be formally achieved through Memoranda of Understanding (MoUs) on port State control, where participating countries undertake to inspect an agreed percentage of the estimated amount of individual foreign ships entering the ports of the participating States. Several MoUs have set inspection goals in the range of 15% to 30% of Convention ships. Nine MoUs (see figure 2) are in existence worldwide, including in Europe, Africa, Asia, Latin America, and the Caribbean. Proper regional co-operation and exchange of boarding results among participating Administrations are an effective enforcement tool and can also reduce the requirement for individual States to board all vessels.

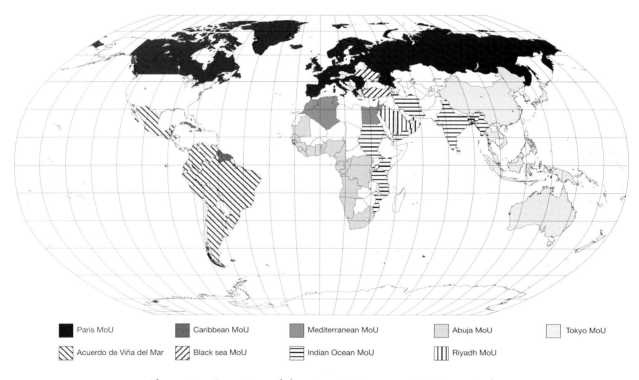

Figure 2 – *Overview of the nine MoUs on port State control*

Part III

Legal aspects

6 Integrating MARPOL into national law

6.1 General

It is assumed that every State Administration will have a legal department or lawyers, which may be attached to its marine administration or to a larger administrative department such as, for example, a Department of Transport. It is further assumed, for the purposes of this manual, that these legal administrators (or lawyers) will have primary responsibility for the legislation that is necessary to implement MARPOL. Whatever the form of the Administration, it must be considered desirable for a single body to be given the overall responsibility for ratification, legislation and implementation. The legal system will vary from State to State, but the principal legal actions necessary for integrating MARPOL into national law and implementation are likely to be as outlined in figure 3 and in the following paragraphs.

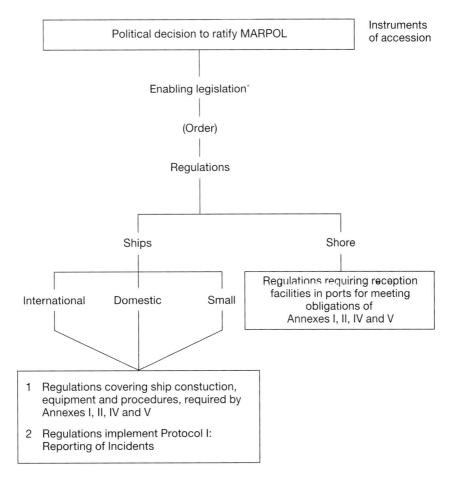

Figure 3 – *Legal actions necessary for integrating MARPOL into national law and for its implementation*

6.2   Instrument of accession

As MARPOL is now in force, the only method of ratification is by accession. In acceding, Governments indicate their acceptance and approval of MARPOL and their readiness to implement its requirements. A document of accession, an "instrument", which may be signed only by the head of State, head of Government or minister for foreign affairs, should be deposited with the Secretary-General of IMO, London (articles 13(2) and IV(2) of MARPOL 73 and the Protocol of 1978 respectively and article 5(2) of the 1997 Protocol). An example, or model, document is given in appendix 1of this manual. An instrument of accession may contain a clause excluding acceptance of one or more of the optional Annexes (Annexes III, IV and V) and the 1997 Protocol (Annex VI).

6.3   Enabling legislation

It is necessary to consider whether existing legislation gives the power through which MARPOL may be integrated into the national legal system. This facility may exist in maritime legislation such as a Mercantile Marine Act, Merchant Shipping Act, or similar legislation. A decision is then necessary on whether any such existing maritime legislation needs amending or whether new legislation, specifically for the purpose of imple-menting MARPOL, is required. It is advisable to look at how other international maritime conventions, such as the International Convention to the Safety of Life at Sea, 1974 (SOLAS), the International Convention on Load Lines, 1966 and the Convention on the International Regulations for Preventing Collisions at Sea, 1972 (COLREG), have been introduced. It is important that implementation of or amendments to MARPOL and associated resolutions and recommendations be permitted. These are frequent and subject in many cases to implementation by an early stated date. An example of enabling legislation covering these points is given in appendix 2 of this manual.

It should be borne in mind that the Convention enters into force three months after the deposit of an instru-ment of accession. The implementing legislation should therefore enter into force not later than at that time. To ensure that this will be the case, the preparation of such legislation has to be initiated well in advance of the accession to the Convention. This timing is obviously particularly important if the implementing legislation is to be adopted by a parliament, congress, etc.

6.4   Order

The legal system of some States may permit regulations to be made directly under the enabling legislation; others require an "order" approved by their Government (e.g., parliament, assembly, congress, legislative assembly, etc.) to bring the various parts of subsidiary legislation into effect. An example of such an order is given in appendix 3 of this manual. It will be seen that the example given would permit the introduction of amendments to MARPOL, and associated resolutions and recommendations, into national regulations rapidly but still under the control of the State Government.

6.5   Regulations

The regulations that compose Annexes I, II, III, IV, V and VI of MARPOL, as amended to date, can, to a large extent, be reproduced as national regulations with very minor changes. Some of the regulations are, however, directed to the State itself, and these are not suitable for straightforward reproduction in national regulations. No attempt is made in this manual to reproduce these MARPOL regulations but specimen suitable national regulations for each Annex are given in appendices 4 to 9, in each of which the appropriate equivalent MARPOL regulations will be recognized. The following comments relate to the regulations under each Annex as numbered in MARPOL.

6.5.1   Annex I (Oil)

Regulation 1. It will be necessary to include some of the definitions given under article 2 of the Convention (e.g., *harmful substance and ship*) and also to define *the sea*.

Regulation 2(1). It will be necessary to change this so that application is only to ships over which a State has jurisdiction (i.e., its own flag ships and others when they are in its waters).

Regulations 3.3, 5.2, 6.2, 6.3.1, 6.3.2, 6.3.4, 7.2, 15.7, 34.7. These regulations lay duties on the Administration and are therefore not suitable as national regulations applying to ships. These duties are covered in chapter 22 of this manual.

Regulation 6.2. Annex I applies to all ships, but surveys and certificates are not required for oil tankers below 150 gross tonnage and other ships below 400 gross tonnage. This regulation requires an Administration to establish appropriate measures for such small ships. National regulations should reflect this requirement (see also chapter 22).

Regulation 7. Small ships (as referenced in regulation 6.2) and ships engaged solely in domestic trade are not required to hold an International Oil Pollution Prevention (IOPP) Certificate. It is likely, in order to control such ships, that a domestic certificate will be required. The national regulations should reflect any such requirement.

Regulation 38. This requires that the Government of each Party undertakes to ensure the provision of reception facilities. This is not suitable as a national regulation and need to be dealt with in a different way. An outline of suitable regulations is given in appendix 10 of this manual, and these are directed at ports and harbours, not to ships.

Additional regulations will be necessary to cover inspection, detention and penalties as required by the articles of MARPOL.

6.5.2   Annex II (Noxious liquid substances in bulk)

In general, the term *Administration* will need to be changed to indicate the marine administration of the State (e.g., the Department of Transport).

Regulation 1. It will be necessary to define *ships, the sea,* etc.

Regulation 2.1. It will be necessary to change this provision so that application is only to ships over which a State has jurisdiction (i.e., its own flag ships and others when they are in its waters).

Regulations 5.2, 6.3, 8.2.1 – 8.2.6, 9.2, 9.3.1 – 9.3.4, 11.2, 13.2.3, 16.1 and 18. These regulations lay duties on Governments or the Administration and are therefore not suitable as national regulations applying to ships. These duties are covered in chapter 22.

Regulation 18. This regulation requires that the Government of each Party undertake to ensure the provision of reception facilities. This regulation is therefore not suitable as part of national regulations applying to ships. An outline of suitable regulations is given in appendix 10; these regulations are directed at ports and harbours.

Additional regulations will be necessary to cover inspection, detention and penalties, as required by the articles of MARPOL.

6.5.3   Annex III (Harmful substances in packaged forms)

The following comments apply to the revised Annex III, which entered into force on 1 January 2010.

Regulation 1.1. This regulation will need to be changed so that application is only to ships over which a State has jurisdiction (i.e., its own flag ships and others when they are in its waters).

Regulation 1.3. This regulation lays a duty on the Government of each Party and is not suitable as part of national regulations applying to ships. The only practical way of applying this regulation is to require compliance with the International Maritime Dangerous Goods (IMDG) Code (implied by a footnote to this regulation); national regulations should reflect this requirement.

Additional regulations will be necessary to cover inspection, detention and penalties as required by the articles of MARPOL.

6.5.4 Annex IV (Sewage)

For this chapter, all references are related to MARPOL Annex IV, as amended by several resolutions up to and including resolution MEPC.200(62). The latter has an entry into force date of 1 January 2013.

Regulation 2. It will be necessary to change this regulation so that application is only to ships over which a State has jurisdiction (i.e., its own flag ships and others when they are in its waters).

Regulations 4.2, 4.3, 6, 12 and 13. These regulations lay duties on the Administration and are therefore not suitable as national regulations applying to ships. These duties are covered in chapter 22.

Regulation 4.2. This regulation requires an Administration to establish appropriate measures for domestic trade ships which are not required to hold an International Sewage Pollution Prevention (ISPP) Certificate or to be surveyed and for ships of less than 400 gross tonnage which are certified to carry 15 persons or less. It is likely, in order to control such ships, that inspection and a domestic certificate will be required: national regulations should reflect any such requirement.

Regulation 12. This regulation requires that the Government of each Party undertake to ensure the provision of reception facilities. It is therefore not suitable as part of national regulations applying to ships. An outline of suitable regulations is given in appendix 10; these regulations are directed at ports and harbours.

Additional regulations will be necessary to cover inspection, detention and penalties as required by the articles of MARPOL.

6.5.5 Annex V (Garbage)

For this chapter, all references are related to the revised MARPOL Annex V which has an entry into force date of 1 January 2013.

Regulation 2. It will be necessary to change this regulation so that application is only to ships over which a State has jurisdiction (i.e., its own flag ships and others when they are in its waters).

Regulation 8. This regulation relates to the provision of reception facilities and requires the Government of each Party to ensure their provision. They are not, therefore, suitable as part of national regulations applying to ships. An outline of suitable regulations is given in appendix 10; these regulations are directed at ports and harbours.

6.5.6 Annex VI (Air pollution and energy efficiency)

In general, the term *Administration* will need to be changed to indicate the marine administration of the State (e.g., the Department of Transport).

Regulation 1. It will be necessary to change this regulation so that application is only to ships over which a State has jurisdiction (i.e., its own-flag ships and others when they are in its waters).

Regulation 3.2. This regulation covers the granting of exemptions for ships participating in trials of emission reduction and control technologies. The national regulations should explain how such exemptions are to be granted and the assessment of the conditions should be attached to those exemptions.

Regulation 3.3.1. This regulation covers certain exemptions applicable to sea-bed mineral activities. Depending on how such activities are to be controlled by national regulations the inclusion of this regulation in national regulations applying to ships may not be appropriate.

Regulation 3.3.2. This covers the use of produced hydrocarbons and hence should be covered instead under national legislation applicable to sea-bed mineral activities.

Regulation 4. This deals with the approval of equivalent alternative means of compliance, particularly in respect of the control of SO_x and particulate matter emissions where the particular means to achieving control are given in the relevant regulation (in contrast, the regulation dealing with NO_x emissions does not specify the means, only the objectives and the survey and certification procedures – which are not open to alternative options). The national regulations should reflect how such equivalent proposals are to be handled.

Regulations 5, 11, 15.3, 18.1, 18.2.5, 18.8.2, 18.10. These regulations lay duties on the Administration and are therefore not suitable as national regulations applying to ships. These duties are covered in chapter 22.

Regulation 5.2. Annex VI applies to all ships, but surveys and certificates are not required for ships below 400 gross tonnage. This regulation requires an Administration to establish appropriate measures for such ships. National regulations should reflect this requirement (see also chapter 22).

Regulation 12.2. Part of this regulation indicates that the leakage of ozone-depleting substances may also be controlled. If it is decided to control such leakages, such a regulation should be given within the national legislation.

Regulation 12.6. This regulation is related to the form of the Ozone-depleting Substances Record Book which should be reflected within the national legislation.

Regulations 13.1.1.2, 13.2.2 (in respect of Tier III engines), 13.5.2.2 (in respect of Tier III engines). These regulations provide that, under certain circumstances, relaxation from the NO_x certification requirements may be accepted. These instances should be addressed within the national legislation.

Regulations 13.1.2.2, 13.1.3. These regulations provide for exemption or exclusion from the requirement for diesel engine NO_x certification under certain circumstances, which should be reflected within the national legislation.

Regulations 13.10, 14.2, 14.8–14.9, 15.4. These regulations relate to functions to be undertaken by IMO and are therefore not suitable as national regulations applying to ships.

Regulation 14.6. This regulation regards the form of the logbook used to record fuel oil changeover operations which should be reflected within the national legislation.

Regulations 15.1 – 15.3. These regulations relate to the designation of certain ports or terminals as requiring certain tankers to be fitted with vapour emission control systems and the provision of the corresponding shore-side facilities. They are not therefore suitable as part of national regulations applying to ships.

Regulation 16.5. This regulation refers to other international instruments and to the development of alternative waste treatment processes and hence is not therefore suitable as part of national regulations applying to ships.

Regulation 16.6.2. This regulation provides that, under the circumstance given, exclusion from the requirements for incinerator certification may be allowed. This case should be addressed within the national legislation.

Regulation 17. This regulation requires that the Government of each Party undertake to ensure the provision of reception facilities. It is therefore not suitable as part of national regulations applying to ships. An outline of suitable regulations is given in appendix 10: these regulations are generally directed at ports and harbours although there is a potential to extend the requirements to include ship-recycling facilities.

Regulation 18.9. This regulation requires the Government of each Party to ensure that appropriate authorities it designates carry out the registration and control of fuel oil suppliers. It is not therefore suitable as part of national regulations applying to ships and rather needs to be implemented through other means appropriate to the wider national legislation. Also applicable are regulations 18.3, 18.4 and 18.5 which effectively relate to the control of fuel oil suppliers together with those aspects of regulation 18.8.1 relating to the provision of the representative sample.

Regulation 18.11. This regulation refers to alternative means by which certain ships may demonstrate compliance with the bunker delivery note requirement. If this option is to be adopted it should be reflected in the national legislation.

Regulation 19. The regulation requires the Government of each Party to apply chapter 4 to all ships of 400 gross tonnage and above, except those ships engaged in voyages within domestic waters or where it chooses to waive the requirements for regulation 20 and 21 subject to the provisions of regulation 19.5.

Regulation 20. This regulation refers to the requirements for new ships and those ships which have undergone major conversion so extensive that it is regarded as a newly constructed ship, as applicable, to have an attained EEDI calculated.

Regulation 21. This regulation refers to the requirements for new ships and those ships which have undergone major conversion so extensive that it is regarded as a newly constructed ship, as applicable, to have the required EEDI determined.

Regulation 22. This regulation refers to the requirement for all ships to keep on board a ship specific Ship Energy Efficiency Management Plan (SEEMP).

Regulation 23. This regulation refers to requirements for the promotion of technical co-operation and transfer of technology relating to improvement of energy efficiency of ships.

Additional regulations will be necessary to cover inspection, detention and penalties as required by the articles of MARPOL.

6.5.7 Regulations implementing Protocol I: reporting of incidents

Regulations are required to give effect to Protocol I of MARPOL. An outline of suitable regulations is given in appendix 11 of this manual. Further consideration to the requirements for reporting incidents involving oil and other harmful substances is given in chapter 16.

6.5.8 Reception facilities

As stated in the foregoing paragraphs (6.5.1, 6.5.2, 6.5.4 6.5.5 and 6.5.6), the regulations requiring the provision of reception facilities are directed at the Government of each Party to MARPOL and are not suitable as national regulations to ships. The means by which Governments ensure the provision of reception facilities at their ports, harbours and terminals will vary but legislation in some form will be necessary. An outline of suitable regulations is given in appendix 10. Further consideration is given to the requirements and provision of reception facilities in chapter 15.

6.5.9 Amendments

The responsibility to keep its national legislation up to date belongs to the marine administration. Keeping track of all amendments to relevant regulations requires a constant input. Relevant guidelines, in particular recommendations, for the correct implementation form a part of this element. An example of a guidance document is the guidance on the timing of replacement of existing certificates as provided in MSC-MEPC.5/Circ.6, as issued on 6 August 2009.

6.5.10 Summary of legal actions

.1   preparation of "instrument of accession";

.2   preparation of enabling legislation;

.3   preparation of "order";

.4   preparation of regulations for implementing Annex I;

.5   preparation of regulations for implementing Annex II;

.6   preparation of regulations for implementing optional Annex III;

.7   preparation of regulations for implementing optional Annex IV;

.8   preparation of regulations for implementing optional Annex V;

.9   preparation of regulations for implementing optional Annex VI;

.10   preparation of regulations for reporting of incidents;

.11   preparation of regulations for provision of reception facilities;

.12   keeping amendments in the national legislation up to date.

7 The legal aspects of enforcement

7.1   What are violations?

It is important that legislation or regulations implementing MARPOL establish the elements of a MARPOL violation such that enforcement personnel or courts are able to ascertain whether clear objective evidence of a violation is present. For instance, it is generally accepted that oil discharged at a rate of 15 ppm will not produce a visible sheen, and that oil discharged at 100 ppm might produce a visible sheen, but only under the most optimum conditions of light and sea state. Therefore, in most every case where a visible sheen is observed trailing a ship and there is no sheen observed ahead of the ship, the ship is operating in violation of MARPOL. Thus, implementing legislation or regulations may establish that evidence of a sheen is presumptively a violation of MARPOL, subject to rebuttal on the part of the ship that it was in compliance with the provisions of the Convention. In this matter, reference is made to resolution MEPC.61(34), Visibility limits of oil discharges of Annex I of MARPOL as an appropriate standard for indicating that a violation has occurred.

While it is recognized that States have different standards of proof under their individual legal systems, in general, States should allow for the reception of a wide variety of credible evidence, including circumstantial evidence, to indicate violations of MARPOL. Some States require, in every case, evidence in the form of chemical analysis, such as "oil fingerprinting", that proves that the discharge came from a particular ship and from no other possible source. This means that other relevant evidence of illegal discharge, such as eyewitness statements, Oil Record Book entries, and so on, is rendered moot. Such an evidentiary requirement on enforcement is burdensome and significantly reduces the ability of the State to enforce the provisions of MARPOL effectively. The gathering, presentation and admitting of evidence for MARPOL violations must be carefully developed by States, where practicable in co-operation with other States, for the effective enforcement of the Convention.

7.2   Sanctions

Article 4(1) of the MARPOL Convention states that "Any violation of the requirements of the present Convention shall be prohibited and sanctions shall be established therefore under the law of the Administration of the ship concerned wherever the violation occurs". MARPOL article 4(4) further provides that sanctions shall be "adequate in severity to discourage violations … and shall be equally severe irrespective of where the violations occur".

The type of sanctions applicable to varying violations under the Convention is a matter for determination of the individual Party and may be a function of several legal, political and economic circumstances, subject to the strictures of MARPOL, article 4(4). Moreover, the approach to sanctions in civil law and in common law jurisdictions may also differ. As sanctions can be very effective as a compliance tool, it is necessary for States to prescribe sanctions that are at least in harmony with applicable systems in neighbouring States or territories so as to avoid the perception that some States have less stringent sanctions than others, as this is one way of insinuating a "safe haven" to the potential polluter. On the other hand, sanctions may take voluntary mitigation efforts and self-reporting into account. Such a progressive system is easier and less expensive to police, and preserves prosecutorial assets for larger cases where substantial harm has occurred.

Flag States should adopt sanctions for those activities that defeat the purposes of the regulations, such as intentional falsification of records required by MARPOL. All States should adopt sanctions for such matters as witness tampering, suborning perjury, interference with law enforcement officials and similar offences.

Sanctions for these types of violations may be deemed criminal and could thereby serve as an important tool in promoting truthfulness in reporting, monitoring, and other regulatory requirements. It is important to note that swift and certain sanctions for violations will have an important deterrent effect. It is also important to note that merely providing for the imposition of sanctions in national legislation will not, on its own, achieve significant benefits. Such sanctions should be supported by effective technical procedures for gathering evidence, as outlined in part V of this manual.

Part IV

Implementing the regulations

8 Implementing the regulations of the Annexes

8.1 General

In chapters 9 to 14 of this manual, a brief description of each of the six Annexes containing the regulations of MARPOL is given. No attempt is made to repeat or explain the regulations, but an explanation is given of what needs to be done by:

.1   the shipbuilder or shipowner (for each type or class of ship) – in respect of design, equipment, construction, procedures, training;

.2   the marine administration – in respect of its own-flag ships, port State duties, coastal State duties; and

.3   the ports – in respect of reception facilities.

Reference is made against each requirement to the applicable regulation of the relevant Annex and to relevant guidance documents produced by IMO. Those responsible for implementing and complying with these regulations must refer to these authoritative and detailed documents.

8.2 Application

It should be understood that, in general, a ship is affected by the provisions of more than one Annex of MARPOL. Annex I, for example, applies to almost all ships, not only oil tankers, with respect to their engine-rooms, because practically all of them use oil as fuel and lubricating oil in their engines. Annex V, concerning garbage, is relevant to all types of ships; so is Annex IV, on sewage and Annex VI, on air pollution. Annexes II and III, however, only apply to ships certified to carry certain types of cargo. Therefore, each ship should be considered to see which Annexes and which parts of each Annex apply.

8.3 Status of this manual

It is emphasized that the Annexes to MARPOL are living documents that develop through time. It is therefore not possible for this manual to reflect texts that are fully updated, and the reader is strongly advised to consult recent updates of the instruments through IMO documents and publications.

9 Implementing Annex I: regulations for the prevention of pollution by oil

9.1   Brief explanation of Annex I

Annex I applies to all ships to which MARPOL applies (see paragraph 3.3). The discharge of oil into the marine environment is prohibited in some areas and severely restricted in others. Ships are required to meet certain equipment and constructional standards and to maintain an Oil Record Book. With the exception of very small ships, a survey is required and, for ships trading internationally, certification in a prescribed form is necessary. Ports are required to provide adequate reception facilities for oil, oily mixtures and oil residue (sludge) to meet the needs of ships using the ports.

In the following sections of this chapter, the basic requirements are outlined for each type and size of ship defined in Annex I. This information is followed by the actions required of the marine administration, and then the action required of the ports.

During the 1990s, it was decided that a major editorial update was needed for MARPOL Annex I to make the Annex more user friendly. Within IMO/MEPC, a correspondence group was established to carry out this major revision and the revised MARPOL Annex I entered into force on 1 January 2007 through resolution MEPC.117(52).

The main improvements are the split in the Annex between MARPOL Annex I – engine room and MARPOL Annex I – cargo area and the addition of definitions related to the delivery date of a ship (definitions under regulation 1.28.1 and further).

The revised MARPOL Annex I contained seven chapters and three appendices, however, meanwhile two additional chapters were adopted which leads to the following layout.

Chapter 1	General	Regulations 1 – 5
Chapter 2	Surveys and certification	Regulations 6 – 11
Chapter 3	Requirements for machinery spaces of all ships	Regulations 12 – 17
Chapter 4	Requirements for the cargo areas of oil tankers	Regulations 18 – 36
Chapter 5	Prevention of oil pollution arising from an oil pollution incident	Regulation 37
Chapter 6	Reception facilities	Regulation 38
Chapter 7	Special requirements for fixed or floating platforms	Regulation 39
Chapter 8	Prevention of pollution during transfer of oil cargo between oil tankers at sea	Regulation 40 – 42
Chapter 9	Special requirements for the use or carriage of oils in the Antarctic area	Regulation 43
Appendix I	Lists of oils	
Appendix II	Form of IOPP Certificate and Supplements	
Appendix III	Form of Oil Record Book	

The following paragraphs outline the requirements to be considered by shipbuilders and shipowners in co-operation with the Administration and recognized organizations (see also chapter 23 of this manual), where applicable, for each type and size of ship indicated. The paragraphs follow the sequence of the layout of the revised Annex I as described above.

The requirements of the regulations in Annex I contain detailed criteria or specifications. Those concerned must study these regulations: no attempt is made here to repeat the detail. It is important to recognize that application dates vary with some of the requirements: these are indicated where still relevant.

9.2 Chapter 1 – General

9.2.1 Definitions and application

After the definitions presented in regulation 1, this chapter contains, in regulation 2, the application of MARPOL Annex I. Unless expressly provided otherwise this Annex applies to all ships. Regulation 2 also identifies to which ships certain requirements of this Annex do not apply, for example to ships other than oil tankers utilized to carry oil in bulk with a certain capacity, requirements for tankers carrying asphalt and requirements where a cargo subject to the provisions of Annex II is carried in a cargo space of an oil tanker.

By resolution MEPC.187(59), four new definitions were added to regulation 1. These additional definitions are on: oil residue (sludge), oil residue (sludge) tank, oily bilge water and oily bilge water holding tank. These new definitions are included in the text of the reworded regulation 12.1 and a new regulation 12.2. Further, in regulations 12.2, 13, 17.2.3, 38.2 and 38.7 the word "sludge" is replaced by the words "oil residue (sludge)" and in 17.2.3 the words "and other oil residues" are deleted.

9.2.2 Ships other than tankers carrying large quantities of oil as cargo

Ships other than tankers fitted with cargo spaces constructed and utilized to carry oil in bulk of an aggregate capacity of 200 m^3 or more have to comply with certain regulations for oil tankers regarding construction and operation (see regulation 2.2 of Annex I), except where the capacity is less than 1,000 m^3. In the latter case, a ship need not to comply with the requirements for slop tanks, oil/water interface detector and oil discharge monitoring and control system, however, the ship needs to comply with the requirements for oil tankers of less than 150 gross tonnage.

9.2.3 Exemptions and waivers, exceptions

Regulation 3 identifies exemptions and waivers for certain ships such as air-cushion vehicles and submarine crafts or for oil tankers solely engaged in specific trades.

Exceptions from discharge requirements are provided for in regulation 4. These exceptions are mainly related to exceptional circumstances like in case to secure the safety of a ship or to save life at sea or in cases involving combating specific pollution incidents.

9.2.4 Equivalents

Based on regulation 5, an Administration may allow any fitting, material, appliance or apparatus to be fitted as an alternative to that required by this Annex, however, this equivalent shall be at least as effective as that required by this Annex. Any such equivalent shall be communicated by the Administration to IMO for circulation to the Parties to MARPOL. An operational method shall never be regarded as an equivalent to a hardware requirement.

9.3 Chapter 2 – Surveys and certification

9.3.1 Surveys

In general every oil tanker of 150 gross tonnage and above, and every other ship of 400 gross tonnage and above shall be subject to the following surveys: initial survey, annual survey, intermediate survey and a renewal survey. An additional survey shall be made whenever any important repairs or renewals are made. Surveys are required to cover all requirements of Annex I (regulation 6), and the condition of the ship and its equipment are to be maintained and may not be changed without prior sanction of the marine administration.

9.3.2 Issue or endorsement of certificate, issue or endorsement of certificate by another Government, form of certificate, duration and validity of certificate

After an initial or renewal survey an International Oil Pollution Prevention (IOPP) Certificate shall be issued either by the Administration or by any persons or organization duly authorized by it. In any case the Administration assumes full responsibility for the certificate (regulation 7). An International Oil Pollution Prevention (IOPP) Certificate is required for ships trading internationally. A certificate is not required for ships in domestic trade but may be required by the marine administration in conjunction with the required surveys.

At the request of the Administration, a Government of a Party to MARPOL may survey a ship and when satisfied with the results, may issue the IOPP Certificate. No IOPP Certificate shall be issued to a ship that is entitled to fly the flag of a State which is not a Party to MARPOL (regulation 8). The IOPP Certificate shall be in the form as given in appendix II to Annex I and shall be at least in English, French or Spanish. If the official language of the issuing country is also used, this shall prevail in case of a dispute or discrepancy (regulation 9).

In general an IOPP Certificate is valid for five years with a certain flexibility as described in regulation 10. A certificate will cease to be valid if relevant surveys are not completed within the period described in regulation 6 or when the certificate is not endorsed in accordance with regulation 6.1.3 or 6.1.4 and upon transfer of the ship to the flag of another State.

9.3.3 Oil tankers of less than 150 gross tonnage and other ships less than 400 gross tonnage

For oil tankers of less than 150 gross tonnage and every other ship of less than 400 gross tonnage, no survey is required, but the ship must comply with appropriate measures established by the marine administration to ensure that the applicable provisions of Annex I are met (regulation 6.2), and the condition of the ship and its equipment must be maintained (regulation 6.4.1). No IOPP Certificate shall be issued to these ships (regulation 7.1).

For regulation 11, port State control on operational requirements, reference is made to paragraph 21.1.4 of this manual.

9.4 Chapter 3 – Requirements for machinery spaces of all ships
(Part A – Construction, Part B – Equipment and Part C – Control of operational discharge of oil)

9.4.1 Chapter 3, Part A – Construction

9.4.1.1 *Sludge tanks and discharge connection*

This part applies to every ship of 400 gross tonnage and above and provides requirements for tanks for oil residues (sludge) (regulation 12) and the standard discharge connection (regulation 13). Regarding regulation 12, it should be noted that the text of the revised Annex I received amendments through resolution MEPC.141(54) adopted on 24 March 2006 and resolution MEPC.187(59) adopted on 17 July 2009. These are important amendments to the original regulation 12 of the revised Annex I (see also paragraph 9.2.1 of this manual).

9.4.1.2 *Fuel tank protection and sludge tank provisions*

Resolution MEPC.141(54) included a new regulation 12A on oil fuel tank protection. The regulation applies to all ships delivered on or after 1 August 2010 with an aggregate oil fuel capacity of 600 m^3 and above. It includes requirements for the protected location of the fuel tanks and performance standards for accidental oil fuel outflow. A maximum capacity limit of 2,500 m^3 per oil fuel tank is included in this new regulation 12A, which also requires Administrations to consider general safety aspects, including the need for maintenance and inspection of wing and double-bottom tanks or spaces, when approving the design and construction of ships which are built in accordance with the regulation.

Regulation 12, as amended, requires that ships shall be provided with sludge tank(s) of adequate capacity taking into account the type of machinery and length of the voyage. Oil residue (sludge) tank(s) shall be provided with a designated pump and shall have no connections to the bilge system, oily bilge water holding tank(s), tank top or oily water separators. Under strict conditions the tank(s) may be fitted with drains that lead to an oily bilge water holding tanks or bilge well (new regulation 12.2).

9.4.1.2 *Provisions for ships of less than 400 gross tonnage*

For ships of less than 400 gross tonnage a holding tank of sufficient capacity might be an option (see paragraph 9.4.2.3).

9.4.2 Chapter 3, Part B – Equipment

9.4.2.1 *Equipment for ships ≥ 400 gross tonnage*

Ships of 400 gross tonnage and above but less than 10,000 gross tonnage require oil filtering equipment such that any oily mixture discharged into the sea aftern passing through the filtering equipment has an oil content not exceeding 15 ppm (regulations 14.1 and 14.6). Ships of 10,000 gross tonnage and over require 15 ppm oil filtering equipment with alarm and automatic stopping device (regulation 14.2 and 14.7).

Ships which are stationary need not be provided with oil filtering equipment but shall be provided with a holding tank adequate for the total retention on board of all oily bilge water (regulation 14.3).

An Administration may waive the requirements for oil filtering equipment for ships engaged exclusively on voyages within special areas, subject to strict conditions being met for holding tanks and reception facilities (regulation 14.5).

9.4.2.2 *Integrated bilge water treatment system (IBTS)*

In this respect attention is drawn to the revised guidelines for systems for handling oily wastes in machinery spaces of ships incorporating guidance; notes for an integrated bilge water treatment system (IBTS) (MEPC.1/Circ.642, as amended by MEPC.1/Circ.676 and MEPC.1/Circ.760). These guidelines are developed for Administrations, shipowners and shipbuilders for consideration in achieving an efficient and effective system for the handling of oily bilge water and oily residues for ships. The following two items received special attention:

.1 To ensure compliance with the provisions on oil residue (sludge) collection and disposal of MARPOL Annex I, the process of regeneration of fuel oil from oil residue (sludge) has been defined and included in the Guidance notes for an integrated bilge water system (IBTS). This guidance is appended to the guidelines.

.2 A recommendation regarding internal drain tanks for oily residue (sludge) and other oily residues, which can only be discharged to the vessel's oily residue (sludge) tanks has been included in order to simplify the required records in the Oil Record Book.

A typical flow diagram of an IBTS is shown in figure 4.

9.4.2.3 *Equipment for ships < 400 gross tonnage*

For ships of less than 400 gross tonnage, the requirements of the flag Administration, to ensure that discharge requirements for these ships are met, are likely to be that the ship is provided with:

.1 the approved oil filtering equipment required by regulation 14.1 (as for ships of 400 gross tonnage and above) and a sludge tank of sufficient capacity for the ship's operational needs; or

.2 a holding tank to retain on board oily mixtures and oil residues, and save-alls or gutters around oil appliances. The holding tank should be of adequate capacity for the ship's operational needs and should be provided with means for transferring the contents of the tank to shore reception facilities.

In figures 5 and 6 examples are given for an oily-water separating arrangement combined with bilge-water settling and for an oily-water separating arrangement, both examples are for small ships only.

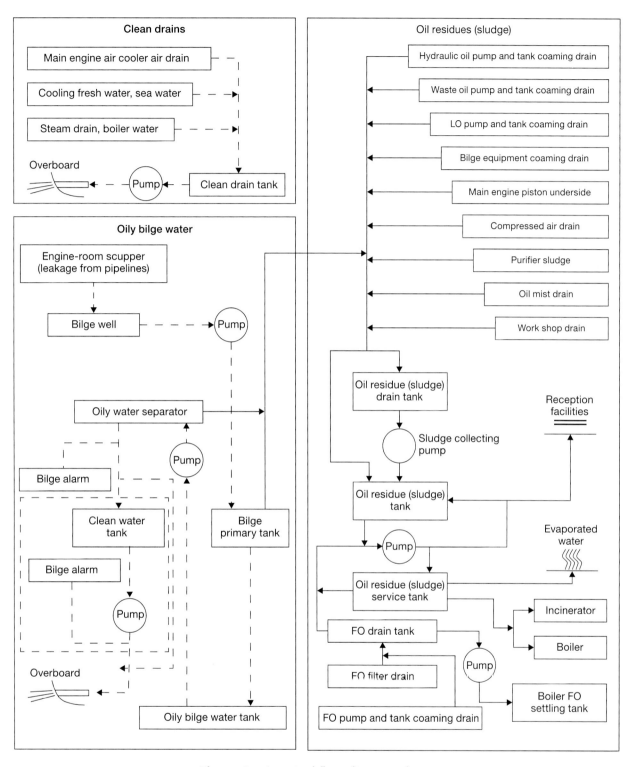

Figure 4 – *A typical flow diagram of IBTS*

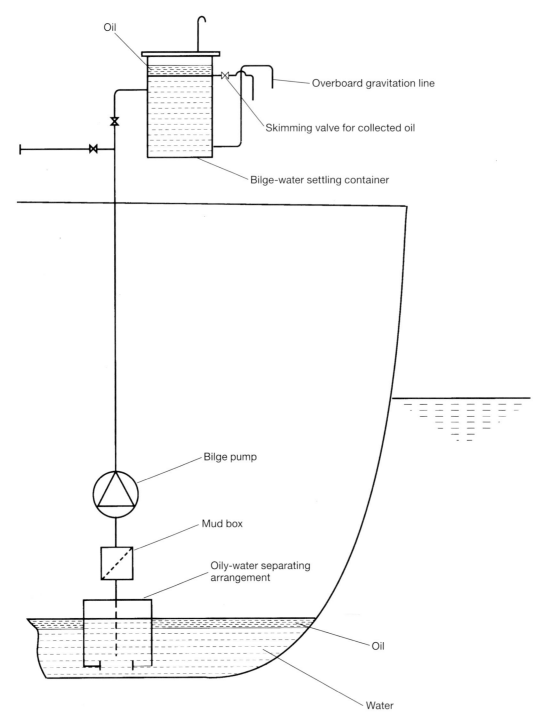

Oil

Overboard gravitation line

Skimming valve for collected oil

Bilge-water settling container

Bilge pump

Mud box

Oily-water separating arrangement

Oil

Water

Notes

1   Bilge-water settling container should have an effective volume (between inlet and outlet) equivalent to 24 h generation of bilge-water.

2   Bilge pump may be mechanical or manual, operating continuously or intermittently.

3   Construction of oily-water separating arrangements is indicated in figure 6.

4   Means of removing residual oil from the surface of the bilge-water and retaining it on board will be necessary.

Figure 5 – *Oily-water separating arrangement combined with bilge-water settling container – for small ships only*

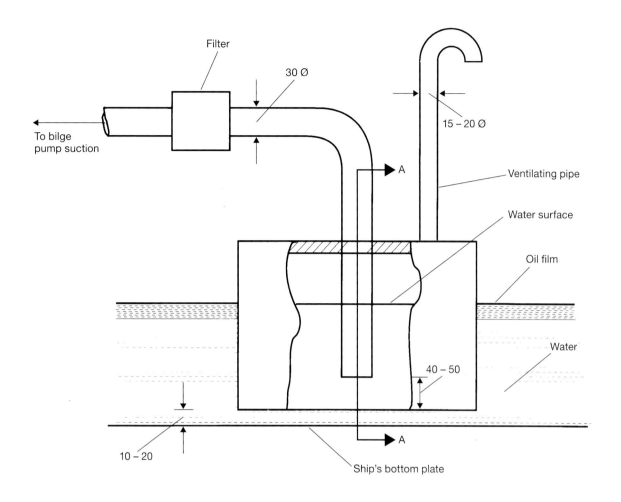

Filter

30 Ø

To bilge
pump suction

15 – 20 Ø

A

Ventilating pipe

Water surface

Oil film

Water

40 – 50

10 – 20

A

Ship's bottom plate

Section through A–A

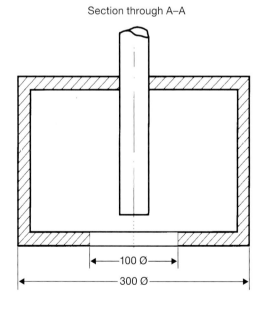

100 Ø

300 Ø

Notes

1   The oily-water separating arrangement should be
located in the ship's machinery space and should be:

.1   attached to the open end of the bilge suction pipe;

.2   positioned aft between two frames on the ship's
centreline;

.3   positioned on an even plane, 10 – 20 mm from the
ship's bottom plate, with the open end of the bilge
suction pipe 50 – 70 mm from the ship's bottom plate.

2   The free area between frames should be bounded
by baffles.

3   The symbol "Ø" refers to diameter; dimensions are in
millimetres.

Figure 6 – *Oily-water separating arrangement – for small ships only*

9.4.3 Chapter 3, Part C – Control of operational discharge of oil

9.4.3.1 *General*

It is important to note that the discharge of oil or oily mixtures into the marine environment is prohibited except when the strict discharge requirements are met. These discharge requirements are divided into those outside and those inside a special area.

It should further be observed that no discharge shall contain chemicals or other substances that are hazardous to the marine environment and that any oil residue that cannot be discharged in compliance with the requirements shall be retained on board for suitable discharge to a shore reception facility.

9.4.3.2 *Special areas (SAs)*

The adoption of stricter discharge requirements for the prevention of marine pollution within special areas is required for technical reasons relating to the oceanographic and ecological conditions of these areas and to their sea traffic.

Special areas under MARPOL Annex I are the Mediterranean Sea, the Baltic Sea, the Black Sea, the Red Sea, the Gulfs area, the Gulf of Aden, the Antarctic area, the North West European waters, the Oman area of the Arabian Sea and the Southern South African waters.

9.4.3.3 *Discharge requirements for ships ≥ 400 gross tonnage outside a SA*

The discharge requirements for ships of 400 gross tonnage and above outside a special area are:

.1   the ship is proceeding *en route*;

.2   the oily mixture is processed through an oil filtering equipment as required in the applicable parts of regulation 14 of Annex I (for ships between 400 and 10,000 gross tonnage regulation 14.6; for ships ≥ 10,000 gross tonnage regulation 14.7);

.3   the oil content of the effluent without dilution does not exceed 15 ppm;

.4   on oil tankers, the oily mixture does not originate from cargo pump-room bilges and is not mixed with oil cargo residues.

9.4.3.4 *Discharge requirements for ships ≥ 400 gross tonnage inside a SA*

The discharge requirements for ships of 400 gross tonnage and above inside a special area are identical to those in 9.4.3.3 with the exception of sub-paragraph .2. Within special areas all ships of 400 gross tonnage and above shall be provided with an oil filtering equipment as required in regulation 14.7, which means an oil filtering equipment with alarm arrangements and an automatic stopping device.

With respect to the Antarctic area, the discharge requirements are even stricter, as any discharge of oil or oily mixtures into the sea from any ship, irrespective of its size, is prohibited.

9.4.3.5 *Discharge requirements for ships < 400 gross tonnage*

For ships of less than 400 gross tonnage the oil or oily mixture shall either be retained on board (see paragraph 9.4.2.3.2) or discharged into the sea (see paragraph 9.4.2.3.1). Where the latter is the case the discharge shall be in accordance with the following requirements:

.1   the ship is proceeding *en route*;

.2   equipment shall be in operation to ensure that the oil content of the effluent without dilution does not exceed 15 ppm;

.3   on oil tankers, the oily mixture does not originate from cargo pump-room bilges and is not mixed with oil cargo residues.

Table 1 – *Control of discharge of oil from machinery spaces of all ships*

Sea area	Ship type and size	Discharge criteria
Anywhere outside a special area	All ships of 400 gross tonnage and above	**No discharge** except when: • the ship is *en route*; • the oily mixture is processed through an oil filtering equipment as required in the applicable parts of regulation 14 of Annex I (for ships between 400 and 10,000 gross tonnage regulation 14.6; for ships $\geq$ 10,000 gross tonnage regulation 14.7); • the oil content of the effluent without dilution does not exceed 15 ppm; • on oil tankers, the oily mixture does not originate from cargo pump-room bilges and is not mixed with oil cargo residues.
Anywhere within a special area	All ships of 400 gross tonnage and above	Same as outside a special area, however, the oil filtering equipment should be provided with alarm arrangements and arrangements that the discharge is automatically stopped when the content of the effluent exceeds 15 ppm.
All areas except the Antarctic area	Ships of less than 400 gross tonnage	**No discharge** except when: Oil or oily mixtures shall be retained on board or discharged into the sea under the following conditions: • the ship is *en route*; • equipment approved by the Administration to ensure that the oil content of the effluent without dilution does not exceed 15 ppm shall be in operation; • on oil tankers, the oily mixture does not originate from cargo pump-room bilges and is not mixed with oil cargo residues.
Antarctic area	All ships irrespective of their size	**No discharge**.
Special areas are the Mediterranean Sea, the Baltic Sea, the Black Sea, the Red Sea, the Gulfs area, the Antarctic area, the North West European waters, the Oman area of the Arabian Sea, the Gulf of Aden and the Southern South African waters.		

9.4.3.6 *Carriage of ballast water in oil fuel tank*

Part C of chapter 3 of Annex I also contains a requirement which prohibits the carriage of ballast water in any oil fuel tank on ships of a certain size delivered after 31 December 1979. However, where there is a need to carry ballast water in an oil fuel tank, such ballast water shall be discharged to a shore reception facility or into the sea. When discharge into the sea takes place this shall be done in compliance with regulation 15 of Annex I (regulations 16.1 and 16.2).

9.4.3.7 *Carriage of oil in a forepeak tank*

For ships of 400 gross tonnage and above with a building contract after 1 January 1982 oil shall not be carried in a forepeak tank or in a tank forward of the collision bulkhead (regulation 16.3).

9.4.3.8 *Requirements for other ships*

With respect to the requirements described in 9.4.3.6 and 9.4.3.7, regulation 16.4 identifies that it is required that all other ships than those mentioned shall comply with the provisions as far as reasonable and practicable.

9.4.3.9 *Oil Record Book*

The Oil Record Book Part 1 – Machinery spaces, is required to be on board of every oil tanker of 150 gross tonnage and above and of every ship of 400 gross tonnage and above. The Oil Record Book (ORB) shall be completed without delay when any of the machinery space operations as identified in regulation 17.2 has taken place. The entry shall be made in at least English, French or Spanish. If the official language of the Administration is also used this shall prevail in case of a dispute or discrepancy. The ORB shall always be promptly available for inspection during a port State control visit.

Over the years it was noted that not all entries in the ORB Part I were completed as they should have been. An inventory showed that there were inconsistencies, mainly of an administrative nature, that raised concerns. The Marine Environment Protection Committee (MEPC) decided to develop guidance in this respect, which was issued under MEPC.1/Circ.736/Rev.2 on 6 October 2011. This guidance is intended to facilitate compliance with MARPOL requirements on board ships by providing advice to crews on how to record the various operations in the ORB Part I by using the correct codes and item numbers in order to ensure a more uniform port State control procedure.

9.5 Chapter 4 – Requirements for the cargo area of oil tankers
(Part A – Construction, Part B – Equipment and Part C – Control of operational discharge of oil)

9.5.1 Chapter 4, Part A – Construction

9.5.1.1 *SBT requirements*

Segregated ballast tank requirements are reflected in regulation 18 which is divided as indicated below:

- oil tankers of 20,000 tonnes deadweight and above delivered after 1 June 1982;

- crude oil tankers of 40,000 tonnes deadweight and above delivered on or before 1 June 1982;

- product carriers of 40,000 tonnes deadweight and above delivered on or before 1 June 1982;

- an oil tanker qualified as a segregated ballast oil tanker;

- oil tankers delivered on or before 1 June 1982 having special ballast arrangements;

- oil tankers of 70,000 tonnes deadweight and above delivered after 31 December 1979; and

- protective location of segregated ballast.

9.5.1.2 *Double hull and double bottom requirements*

Requirements for double hull (DH) and double bottom (DB) for oil tankers delivered on or after 6 July 1996 are explained in regulation 19. The DH and DB requirements for oil tankers delivered before 6 July 1996 are given in regulation 20. The latter contains a phase-out schedule ending in such a way that oil tankers delivered in 1984 or later shall comply with the requirements not later than the anniversary of delivery date in 2010. The possibility for an Administration to suspend the DB/DH requirements for an oil tanker is provided for via the so-called Condition Assessment Scheme (CAS) as referenced in regulation 20.7.

Regulation 21 describes requirements for the prevention of oil pollution from oil tankers carrying heavy grade oil (HGO) as cargo. A main item of this regulation is the DH/DB requirements applicable when an oil tanker is carrying HGO's.

Also related to the DB/DH requirements is regulation 22 which describes pump-room bottom protection. This regulation applies to oil tankers of 5000 tonnes deadweight and above constructed on or after 1 January 2007.

9.5.1.3 *Design and construction requirements*

Other items related to the ship design and construction covered in this part of chapter 4 are:

- accidental oil outflow performance (regulation 23);

- damage assumptions (regulation 24);

- hypothetical outflow of oil (regulation 25);

- limitations of size and arrangement of cargo tanks (regulation 26);

- intact stability (regulation 27); and

- subdivision and damage stability regulation 28).

9.5.1.4 *Slop tanks*

Slop tanks are required on every oil tanker of 150 gross tonnage and above. For oil tankers delivered on or before 31 December 1979, any cargo tank may be designated as a slop tank. The slop tank or the combination of slop tanks shall have a capacity that is necessary to retain on board the slops generated by, for instance, tank washings. The total capacity shall not be less than 3% of the oil-carrying capacity of the oil tanker, however, the Administration may accept a reduction as reflected in regulation 29.2.3.

9.5.1.5 *Pumping, piping and discharge arrangements*

Pumping, piping and discharge arrangements for different oil tankers, oil tankers of different sizes and oil tankers with different delivery dates are reflected in regulation 30.

9.5.2   Chapter 4, Part B – Equipment

9.5.2.1 *Discharge monitoring and control systems*

Oil discharge monitoring and control systems approved by the Administration are mandatory on board oil tankers of 150 gross tonnage and above. The system shall be fitted with a recording device to provide a continuous record of the discharge. The system shall be in operation when there is any discharge of effluent into the sea and shall ensure that the discharge stops automatically when the permitted discharge rate is exceeded.

Failure of the system shall stop the discharge. In the event of failure of the system, a manually operated alternative may be used but repair shall be carried out as soon as possible. In such case the PSC authority may allow the tanker to undertake one ballast voyage before proceeding to the repair port.

With respect to oil discharge monitoring and control systems reference is made to different guidelines relevant for oil tankers with different building dates as indicated in the schedule below.

Table 2 – *Cross reference for discharge monitoring and control systems*

Building date	Relevant guideline
Prior to 2 October 1986	Resolution A.393(X)
On or after 2 October 1986	Resolution A.586(14)
On or after 1 January 2005	Resolution MEPC.108(49)

9.5.2.2 *Oil/water interface detector*

Another part of equipment is the oil/water interface detector (regulation 32), which needs to comply with the specification of resolution MEPC.5(XIII).

The final regulation of this part, regulation 33, prescribes the crude oil washing (COW) requirements. Every crude oil tanker of 20,000 tonnes deadweight and above delivered after 1 June 1982 shall be fitted with a COW system approved by the Administration. More information on COW is provided in Assembly resolution A.446(XI), which has been amended by A.497(XII) and A.897(21).

9.5.3   Chapter 4, Part C – Control of operational discharge of oil

9.5.3.1 *General*

For oil tankers, it is important to note that the discharge of oil or oily mixtures into the marine environment is prohibited except when the strict discharge requirements are met. It should further be observed that no discharge shall contain chemicals or other substances that are hazardous to the marine environment and that any oil residue that cannot be discharged in compliance with the requirements shall be retained on board for suitable discharge to a shore reception facility.

9.5.3.2 *Special areas*

As indicated in paragraph 9.4.3.1, the discharge requirements are divided into those inside and those outside a special area. Special areas under Annex I are the sea areas as indicated in paragraph 9.4.3.2.

9.5.3.3 Discharge requirements outside a SA

The discharge requirements for any oil or oily mixtures from the cargo area of any oil tanker outside a special area are:

- – the tanker is more than 50 nautical miles from the nearest land;

- – the tanker is proceeding *en route*;

- – the instantaneous rate of discharge of oil content does not exceed 30 L per nautical mile;

- – the total quantity of discharge shall not exceed the quantity as reflected in regulation 34.1.5; and

- – the tanker has in operation the oil discharge monitoring and control system and a slop tank arrangement as required in regulations 29 and 31.

These requirements do not apply to the discharge of clean or segregated ballast.

9.5.3.4 Discharge requirements inside a SA

With respect to the discharge of oil or oily mixtures from the cargo area of any oil tanker within Special Areas it should be noted that there is a general prohibition of any of such discharges.

9.5.3.5 Discharge requirements for oil tankers of less than 150 gross tonnage

The requirements for slop tank(s), oil discharge monitoring and control systems and the oil/water interface detector do not apply to these oil tankers, however, the ship shall be so designed and equipped in such way that all oil and oily mixtures can be retained on board with subsequent discharge to shore reception facilities. The total amount of oil and water shall be kept on board in a storage tank unless adequate arrangements are made to ensure that any discharge shall be effectively monitored to ensure compliance with the applicable requirements.

Table 3 – *Control of discharge of oil from cargo tank areas of oil tankers*

Sea area	Ship type and size	Discharge criteria
Within 50 nautical miles from land	Any oil tanker	**No discharge** except for clean or segregated ballast.
Anywhere outside a special area except within 50 nautical miles from land	Any oil tanker	**No discharge** except when: • the ship is *en route*; • the instantaneous rate of discharge of oil content does not exceed 30 L per nautical mile; • the total quantity of discharge shall not exceed the quantity as reflected in regulation 34.1.5 • the tanker has in operation an oil discharge monitoring and control system and a slop tank arrangement as required in regulations 29 and 31. These requirements do not apply to the discharge of clean or segregated ballast.
Anywhere within a special area	Any oil tanker	**No discharge** except for clean or segregated ballast.
All areas except the Antarctic area	Oil tankers of less than 150 gross tonnage	In principle, **no discharge** and retention on board of all oil/oily mixtures in a storage tank and discharge to a shore reception facility, unless adequate arrangements are made so that the discharge is effectively monitored to ensure compliance with the relevant requirements.
Antarctic area	All ships irrespective of their size	**No discharge**.
Special areas are the Mediterranean Sea, the Baltic Sea, the Black Sea, the Red Sea, the Gulfs area, the Antarctic area, the North West European waters, the Oman area of the Arabian Sea, the Gulf of Aden and the Southern South African waters.		

9.5.3.6 *Oil Record Book*

The Oil Record Book Part II – Cargo /ballast operations, is required to be on board of every oil tanker of 150 gross tonnage and above. The Oil Record Book (ORB) shall be completed without delay when any of the operations as identified in regulation 36.2 has taken place. The entry shall be made at least in English, French or Spanish. If the official language of the Administration is also used this shall prevail in case of a dispute or discrepancy. The ORB shall always be promptly available for inspection during a port State control visit.

9.6   Chapter 5 – Prevention of pollution arising from an oil pollution incident

9.6.1   Shipboard oil pollution emergency plan

Every oil tanker of 150 gross tonnage and above and every ship other than an oil tanker of 400 gross tonnage and above shall have on board a shipboard oil pollution emergency plan approved by the Administration. The plan shall be developed in accordance with resolution MEPC.54(32) which is amended by MEPC.86(44) and should be written in the working language of the master and officers. The aim of the plan is that all information necessary to reduce or control the discharge after an oil pollution incident is readily available, such as the procedure for reporting, all contact details and actions to be taken.

9.7   Chapter 6 – Reception facilities

Regulation 38 of Annex I states that the Government of each Party to MARPOL undertakes to ensure the provision of adequate reception facilities for oily mixtures and residues in all ports. These reception facilities are necessary for implementing the Annex.

The method by which a government ensures this provision is covered in paragraph 6.5.8 and chapter 15 of this manual.

9.8   Chapter 7 – Special requirements for fixed or floating platforms

When fixed or floating platforms are engaged in exploration, exploitation and processing they must comply with the requirements for ships other than tankers of 400 gross tonnage and above except that they shall be equipped as far as practicable with sludge tanks and filtering equipment. The discharge of oily mixtures is prohibited except when the oil content does not exceed 15 ppm.

9.9   Chapter 8 – Prevention of pollution during transfer of oil cargo between oil tankers at sea

On 17 July 2009, the MEPC adopted a new chapter 8 via resolution MEPC.186(59) to include new regulations 40 to 42 in the text of Annex I. These regulations which concern the scope of application, general rules on safety and environmental protection and notification related to the ship to ship transfer of cargo oil entered into force on 1 January 2011. It should be noted that the regulations on transfer of oil cargo between oil tankers at sea do not apply to bunkering operations.

9.10   Chapter 9 – Special requirements for the use or carriage of oils in the Antarctic area

On 26 March 2010, the MEPC adopted a new chapter 9 via resolution MEPC.189(60) to include new regulation 43 in the text of Annex I. These requirements regulate the prohibition of the carriage of certain types of oil in bulk as cargo and the prohibition of the carriage and use of these types of oil as fuel in the Antarctic area entered into force on 1 August 2011.

9.11 Records and documents on board

Based on the requirements in Annex I the following records and documents are required to be carried on board of an oil tanker (OT) or by a ship other than an oil tanker (non OT);

.1 Oil Record Books, Part I (OT and non OT; regulation 17) and Part II (OT, regulation 36)

.2 Loading and Damage Stability Information Book (OT; regulation 28.5)

.3 ODM Operational Manual (OT; regulation 31.4)

.4 COW Operations and Equipment Manual (OT; regulation 33) (if relevant)

.5 CBT Operation Manual (OT; regulation 18.8) (if relevant)

.6 Instructional or operational Manual for oily-water separator and oil filtering equipment (OT and non OT; regulation 14)

.7 Shipboard Oil Pollution Emergency Plan (OT and non OT; regulation 37).

9.12 Actions by the marine administration

9.12.1 The marine administration is required to carry out the following in order to implement Annex I:

.1 notify IMO of any exemptions granted under regulation 3;

.2 notify IMO of any equivalent fittings, material, appliance or apparatus allowed as an alternative to that required by the regulations (regulation 5);

.3 carry out initial, periodical and intermediate surveys (regulation 6);

.4 establish measures for ships not subject to surveys under the Annex (regulation 6);

.5 institute unscheduled inspections (regulation 6);

.6 delegate surveys if necessary (regulation 6.3.1) (see also chapter 23 of this manual);

.7 issue certificates following surveys, in the prescribed format (regulation 7);

.8 investigate possible violations of the discharge requirements (regulation 15.7);

.9 approve COW systems (regulation 33), installations and manuals (regulation 35);

.10 approve oil content meters (regulation 31);

.11 approve CBT Operation Manuals (regulation 18.8.4);

.12 consider equipment waivers (regulation 3.4 and regulation 14.5);

.13 approve and agree special ballast arrangements (regulation 18.10);

.14 approve oil discharge monitoring and control systems and instructions (regulation 31);

.15 approve oil/water interface detectors (regulation 32);

.16 approve oil filtering equipment (regulation 14);

.17 develop or approve Oil Record Books for oil tankers of less than 150 gross tonnage (regulation 36.9) and for drilling rigs and platforms (regulation 39.2.2);

.18 approve shipboard oil pollution emergency plans (regulation 37); and

.19 consider and approve ship equipment and construction to meet the requirements of the Annex (in conjunction with surveys).

9.12.2 Guidance documents, recommendations and specifications have been developed by IMO and are listed in chapter 25. These should be used by marine administrations in conducting the duties outlined in paragraph 9.12.1.

9.13 Action by the ports

Actions by ports are mainly related to reception facilities, as mentioned in paragraph 9.7 above. Through the Government of each Party to the Convention, it should be ensured that reception facilities are provided in oil loading terminals, repair ports and in other ports in which ships have oily residues and oily mixtures to discharge. IMO has developed guidelines for ensuring the adequacy of port waste reception facilities to assist ports to comply with the requirements as reflected in regulation 38 of Annex I. These guidelines were adopted via resolution MEPC.83(44).

9.14 Summary of actions and requirements for implementation of Annex I

9.14.1 Ship operators or owners should ensure that every ship:

 .1   has been equipped to Annex I requirements;

 .2   has been surveyed if over 400 gross tonnage or, if a tanker, over 150 gross tonnage;

 .3   has an appropriate certificate (IOPP or domestic);

 .4   has an Oil Record Book; and

 .5   has a crew instructed and trained to comply with the discharge criteria.

9.14.2 Ports should have adequate reception facilities.

9.14.3 The marine administration is required to take the actions outlined in paragraph 9.12.

10 Implementing Annex II: regulations for the control of pollution by noxious liquid substances

10.1   Brief explanation of Annex II

Annex II applies to the carriage in bulk of all noxious liquid substances (NLS) except oil (as defined in Annex I to the Convention). Liquid substances posing a threat of harm to the marine environment are divided into three categories, Category X, Y and Z. Category X presenting a major hazard if discharged into the marine environment, Category Y presenting a hazard and Category Z presenting a minor hazard.

A fourth category, other substances (OS), identifies those liquid substances that, at present, are considered to present no hazard to the marine environment.

The categorization of NLS is based on guidelines reflected in appendix 1 to Annex II. Each category has specific requirements when it comes to carriage, discharge and administration.

The noxious liquid substances (NLS) concerned might, in addition to presenting a pollution hazard, also present a safety hazard. Categories X, Y and Z substances with a safety hazard are listed as such in Chapter 17 of the International Code for the Construction and Equipment of Ships carrying dangerous Chemicals in Bulk (IBC Code), while some Category Z substances that do not present a safety hazard and substances not posing a threat to the marine environment (OS) are listed as such in Chapter 18 of the IBC Code. To avoid "double listing" all liquid substances assessed and therefore allowed to be carried at sea are listed in either Chapter 17 or Chapter 18 of the IBC Code. When listed in Chapter 17, the NLS is identified as presenting a pollution hazard (P) or a safety and pollution hazard (S/P). No noxious liquid substances are listed in Annex II itself.

Annex II prohibits the discharge into the sea of any effluent containing substances falling into Category X, Y or Z except when the discharge is made under conditions which are specified in detail for each category. Annex II requires that every ship be provided with pumping and piping arrangements to ensure that each tank designated for the carriage of NLS does not retain, after unloading, a quantity of residue in excess of the quantity given in the Annex. For each tank intended for the carriage of NLS an assessment of the residue quantity has to be made. Only when the residue quantity, as assessed, is less than the quantity prescribed by the Annex for that particular Category may a tank be approved for the carriage of the specific NLS Category.

In addition to the conditions referred to above, an important requirement contained in Annex II is that the discharge operations of certain cargo residues and certain cleaning and ventilation operations may only be carried out in accordance with approved procedures and arrangements based upon standards developed by IMO. To enable compliance with this requirement, a Procedures and Arrangements (P & A) Manual is required which contains all particulars of the ship's equipment and arrangements, operational procedures for cargo unloading and tank stripping, procedures for the discharge of cargo residues, for tank washing, for slops collection, and for ballasting and de-ballasting as may be applicable to the substances the ship is certified to carry. Following the procedures set out in this ship's P & A Manual will ensure that the ship complies with all relevant requirements of Annex II.

In the early nineties, it was recognized that Annex II was complex due to the many references and was not in line with current technology. In addition, the Earth Summit on Sustainable Development (ESSD) led to a globally harmonized system (GHS) for supply and use of all dangerous products, including chemicals. These developments led to a major revision of Annex II, which was adopted by resolution MEPC.118(52) and entered into force on 1 January 2007.

10.2   Basic requirements of Annex II

In the following sections of this chapter, the basic requirements for implementing Annex II are outlined. The requirements involve action by producers of liquid substances, by shippers, by ship operators or owners, by ports and by the marine administration. The requirements of the regulations are complex and many are detailed. Those concerned with the carriage of these liquid substances must study these regulations: no attempt is made here to repeat the detail but reference is made to the relevant regulations or documents, where appropriate.

The revised MARPOL Annex II contains eight chapters and seven appendices which lead to the following layout.

Chapter 1	General	Regulations 1 – 5
Chapter 2	Categorization of noxious liquid substances	Regulation 6
Chapter 3	Surveys and certification	Regulations 7 – 10
Chapter 4	Design, construction, arrangement and equipment	Regulations 11 – 12
Chapter 5	Operational discharges of residues of NLS	Regulation 13 – 15
Chapter 6	Measures of control by port States	Regulation 16
Chapter 7	Prevention of pollution arising from an incident involving noxious liquid substances	Regulation 17
Chapter 8	Reception facilities	Regulation 18
Appendix 1	Guidelines for the categorization of NLS	
Appendix 2	Form of Cargo Record Book for ships carrying NLS in bulk	
Appendix 3	Form of International Pollution Prevention Certificate for the Carriage of noxious liquid substances in bulk (NLS Certificate)	
Appendix 4	Standard format for the Procedures and Arrangements Manual	
Appendix 5	Assessment of residue quantities in cargo tanks, pumps and associated piping	
Appendix 6	Prewash procedures	
Appendix 7	Ventilation procedures	

10.3   Chapter 1 – General

10.3.1 Definitions and application

Next to the definitions in regulation 1, which for easy reference are put in alphabetical order, this chapter contains, in regulation 2, the application of MARPOL Annex II. Unless expressly provided otherwise, this Annex applies to all ships certified to carry NLS in bulk. In regulation 2 it is also identified that, where a cargo, subject to the provisions of Annex I, is carried in a cargo space of an NLS tanker, the appropriate requirements of Annex I apply for that cargo.

10.3.2 Exceptions, exemptions and equivalents

Exceptions from discharge requirements are provided for in regulation 3. Exceptional circumstances might include the need to secure the safety of a ship or save life at sea or combat specific pollution incidents.

Regulation 4 describes exemptions. The most important one is related to the carriage of vegetable oils. For specific vegetable oils, which are identified by a footnote in the IBC Code, the carriage requirements for a Ship Type 2 may be waived but only when all of the requirements in regulation 4.1.3 are met.

Based on regulation 5, an Administration may allow any fitting, material, appliance or apparatus to be fitted as an alternative to that required by this Annex, however, this equivalent shall be at least as effective as that required by this Annex. Any such equivalent shall be communicated to IMO for circulation to the Parties to MARPOL. An operational method shall never be regarded as an equivalent to a hardware requirement.

The requirements for the construction and equipment for liquefied gas carriers certified to carry NLS are also provided in regulation 5.

10.4   Chapter 2 – Categorization of noxious liquid substances

10.4.1 Categorization before transport

Any liquid substance, other than oil, offered for transport in bulk shall be evaluated and classified into one of the four categories, X, Y, Z or OS (see paragraph 10.1). It is therefore necessary that the shipper should first check whether the product is listed in either chapter 17 or 18 (or in chapter 19 – Index of products carried in bulk) of the IBC Code. It is important to note that the product name, which is the name in column a of chapter 17 or the name in chapter 18, shall be used in the shipping document. Any additional name may be included in brackets after the product name.

10.4.2 Provisional assessment

If the product is not yet reflected in the IBC Code, a provisional assessment should be established. The provisional assessment is an agreement among the producing or shipping country, the flag State and the receiving State. Guidelines for the provisional assessment of liquid substances transported in bulk are given in MEPC.1/Circ.512. Those products already provisionally assessed can be found in the latest edition of MEPC.2/Circular (issued every year on 17 December) or on the IMO website. It should be noted that a provisional assessment ceases to be valid three years after its first publication in the MEPC.2/Circular. Carriage of a product not yet (provisionally) assessed is prohibited.

10.5   Chapter 3 – Surveys and certification

10.5.1 Surveys

In general every ship carrying NLS in bulk shall be subject to the following surveys:

Initial survey, annual survey, intermediate survey and renewal survey. An additional survey shall be made whenever any important repairs or renewals are made. Surveys are required to cover all applicable requirements of Annex II (regulation 8), and the condition of the ship and its equipment are to be maintained and may not be changed without prior sanction of the marine administration.

10.5.2 Issue or endorsement of certificate

After an initial or renewal survey, an International Pollution Prevention Certificate for the Carriage of Noxious Liquid Substances in bulk (NLS Certificate) shall be issued either by the Administration or by any persons or organization duly authorized by it. In every case the Administration assumes full responsibility for the certificate (regulation 9). No NLS Certificate shall be issued to a ship that flies the flag of a State which is not a Party to MARPOL (regulation 9.3.4). The NLS Certificate shall be in the form as given in appendix 3 to Annex II and shall be at least in English, French or Spanish. If the official language of the Administration is also used this shall prevail in case of a dispute or discrepancy (regulation 9.4).

10.5.3 Duration and validity of certificate

In general, an NLS Certificate is valid for five years with a certain flexibility as described in regulation 10. A certificate will cease to be valid if relevant surveys are not completed within the period described in regulation 8, when the certificate is not endorsed in accordance with regulation 8.1.3 or 8.1.4 or upon transfer of the ship to the flag of another State.

10.5.4 NLS Certificate and Certificate of Fitness

Any chemical tanker which has been surveyed under the provisions of the IBC Code shall also be deemed to comply with the requirements under Annex II of MARPOL. This means that the NLS Certificate is incorporated in the International Certificate of Fitness for the Carriage of Dangerous Chemicals in Bulk issued under the requirements of the IBC Code.

10.6   Chapter 4 – Design, construction, arrangement and equipment

10.6.1 Design, construction, arrangement and equipment

All ships certified to carry those NLS identified in chapter 17 of the IBC Code shall be in compliance with the relevant requirements in regulation 11.1 of Annex II so as to minimize the accidental discharge into the sea of such substances. In the case that NLS identified in chapter 17 of the IBC Code are carried on ships other than chemical tankers, for instance in limited quantities on offshore support vessels or for certain vegetable oils in general dry cargo ships, the marine administration shall establish appropriate measures based on the relevant guidelines established by IMO.

10.6.2 Pumping, piping, unloading arrangements

Annex II is based on the principle that no ship is allowed to proceed to sea with more than the maximum allowed residual quantity of a product in the tank and associated piping after the tank has been unloaded. The maximum residual quantity depends on the Category of the product and the construction date of the ship (see table 4).

Table 4 – *Maximum allowed residual quantity per tanks and its associated piping*

Category	New ships*	IBC†	BCH‡	Other ships§
X	75 L	100 L + 50 L tolerance	300 L + 50 L tolerance	No carriage
Y	75 L	100 L + 50 L tolerance	300 L + 50 L tolerance	No carriage
Z	75 L	300 L + 50 L tolerance	900 L + 50 L tolerance	New ships: 75 L Existing ships: empty tanks to the most practicable extent
OS	unrestricted	unrestricted	unrestricted	unrestricted

* New ships means ship constructed on or after 1 January 2007.

† IBC means ship constructed on or after 1 July 1986 but before 1 January 2007.

‡ BCH means constructed before 1 July 1986.

§ Other ships means ships other than a NLS or Chemical tanker; new ships means constructed on or after 1 January 2007.

The residual quantity has to be established via a pumping performance test as set out in appendix 5 to Annex II. Water shall be used as a test medium.

10.6.3 Underwater discharge outlet

Ships certified to carry NLS shall have underwater discharge outlet(s) in compliance with regulation 12.9. For ships, constructed before 1 January 2007 and only certified to carry NLS in Category Z the underwater discharge outlet is not mandatory.

10.6.4 Slop tanks

Annex II does not require the fitting of a dedicated slop tank, however, slop tanks may be needed for certain washing procedures. Under the provisions of Annex II any cargo tank may be used as a slop tank.

10.7 Chapter 5 – Operational discharges of residues of NLS

10.7.1 Control of discharges

It is important to note that any discharge into the sea of products of Category X, Y or Z is prohibited unless such discharges are made in compliance with the requirements in regulation 13 of Annex II.

Before any discharge into the sea takes place the following operation shall be carried out:

Table 5 – *General operational requirement per pollution category to achieve no more than the maximum allowed residual quantity per tank and its associated piping*

Category	Operation
X	Prewash
Y	High viscosity and solidifying products → prewash
	Non high viscosity and non-solidifying products → efficient stripping
Z	Efficient stripping

Any subsequent discharge of water added to the tank shall take place under the following conditions:

- the ship is *en route*;
- the ship has a speed of at least 7 knots (non self-propelled 4 knots);
- the discharge is made below the waterline;
- the distance from the nearest land is not less than 12 nautical miles; and
- the depth of the water is not less than 25 m.

10.7.2 Exemption for a prewash (regulation 13.4)

On request of the ship's master, an exemption for a prewash may be granted when:

- the unloaded tank will be reloaded with the same or a compatible cargo;
- the unloaded tank will neither be washed nor ballasted at sea and the prewash will take place in another port – confirmation of availability of shore reception facilities in that port shall be available in writing; or
- the cargo residue will be removed by ventilation.

10.7.3 Use of cleaning agents or additives

When a washing medium other than water is used, it is called a "cleaning agent". After a cleaning agent has been used the tank should be treated as if the cleaning agent had been carried as the last cargo.

When a small amount of detergent products is added to the cleaning water, it is called a "cleaning additive". Cleaning additives are only allowed to be used when they have been evaluated to ascertain whether they meet the specified requirements (regulation 13.5.2). Guidelines for the evaluation of cleaning additives are provided in MEPC.1/Circ.590 (Revised tank cleaning additives guidance note and reporting form).

10.7.4 Procedures and Arrangements Manual

Every ship certified to carry NLS shall have on board a Procedures and Arrangements (P & A) Manual approved by the marine administration. The Manual should be seen as the passport of the ship, with all design, equipment and operations explained, and forms a part of the NLS Certificate (no Manual means no certificate). The Manual is important for use by ship's officers as well as by port State inspectors.

10.7.5 Cargo Record Book (CRB)

The CRB is required to be on board every ship certified to carry NLS. The CRB shall be completed without delay after completion of any operation specified in appendix 2 to Annex II. The entry shall be made in English, French or Spanish. If the official language of the Administration is also used this shall prevail in case of a dispute or discrepancy. The CRB shall always be promptly available for inspection during a port State control visit.

10.8   Chapter 6 – Measures of control by port States

10.8.1 Appointed or nominated surveyors

In contrast to the provisions in other Annexes to MARPOL, under Annex II, the Government of each Party to the Convention shall appoint or authorize surveyors to implement regulation 16. The main purpose is that the surveyor shall witness any prewash of a tank unloaded from an NLS cargo in Category X and make an appropriate entry into the CRB. Also any other operation the surveyor has verified shall be followed by an appropriate entry in the CRB.

10.8.2 PSC on operational requirements

For regulation 16.9, port State control on operational requirements, reference is made to paragraph 21.1.4 of this manual.

10.9   Chapter 7 – Prevention of pollution arising from an incident involving NLS

10.9.1 Shipboard marine pollution emergency plan for NLS

Every ship of 150 gross tonnage and above certified to carry NLS shall have on board a shipboard marine pollution emergency plan for NLS approved by the Administration. The plan shall be developed in accordance with resolution MEPC.85 (44) as amended by MEPC.137(53). The plan shall be written in the working language of or languages understood by the master and officers. The aim of the plan is that all information necessary to reduce or control the discharge after an pollution incident be readily available, such as the procedure for reporting, all contact details and actions to be taken.

If the ship also requires an oil pollution emergency plan under Annex I (see paragraph 9.6.1 of this manual), the two plans may be combined and should be renamed to "Shipboard marine pollution emergency plan".

10.10 Chapter 8 – Reception facilities

Regulation 18 of Annex II states that the Government of each Party to MARPOL undertakes to ensure the provision of adequate reception facilities for the needs of ships using its ports, terminals or repair ports. The method by which a Government ensures this provision is covered in 6.5.8 and chapter 15 of this manual.

10.11 Summary of actions and requirements for implementation of Annex II

10.11.1 Producers and shippers should ensure that, before any liquid substance is offered for shipment in bulk, its characteristics are known and, if it is not an oil, that in accordance with regulation 6:

    .1   it has been evaluated and assessed and is listed in the IBC Code; or

    .2   if it is not listed, it has been provisionally assessed by IMO; or

    .3   if not listed or provisionally assessed by IMO, the guidelines for provisional assessment are followed and conditions for its carriage are established with the relevant marine administration (MEPC.1/Circ.512);

    .4   if one of the sub paragraphs above is not complied with, the cargo shall be refused for carriage; and

    .5   the ship to be used is suitable and appropriately certified.

10.11.2 Ship operators or owners should ensure that any ship to be used for the transport of noxious liquid substances in bulk is suitable and:

- **.1** has been surveyed and tests have been carried out to the Annex II requirements;

- **.2** has an appropriate certificate (NLS or CoF) which includes the substance to be carried;

- **.3** has an approved Procedures and Arrangements Manual;

- **.4** has a Cargo Record Book;

- **.5** has a properly trained crew.

10.11.3 Ports should not accept cargoes of noxious liquid substances unless:

- **.1** arrangements are made for adequate reception facilities for any tank washings or residues that must be discharged in compliance with Annex II;

- **.2** the terminal has suitable arrangements to facilitate the stripping of ships' cargo tanks; and

- **.3** the unloading arrangements do not require cargo hoses or piping to be drained back to the ship (regulation 18.4).

10.11.4 The marine administration is required to take the following actions:

- **.1** arrange provisional assessments of substances as the producing or shipping State, the flag State or the port State (regulation 6.3);

- **.2** appoint surveyors and carry out control measures as a port State (regulation 16);

- **.3** notify IMO of any equivalent fittings, material, appliance or apparatus allowed as an alternative to that required by the regulations (regulation 5);

- **.4** approve ships' Procedures and Arrangements Manuals (regulation 14);

- **.5** approve pumping and piping systems via their testing for assessment of amount of residue (regulation 12 and Appendix 5);

- **.6** consider conditions for permitting waivers of stripping for ships on restricted voyages (regulation 11);

- **.7** carry out initial, periodical and intermediate surveys (regulation 8);

- **.8** delegate surveys, if necessary (regulation 8);

- **.9** issue certificates in the prescribed format following survey (regulation 9); and

- **.10** arrange for adequate reception facilities for residues of NLS.

10.12 Guidance

Guidance documents, recommendations and specifications have been developed by IMO and are listed in chapter 25. These should be used by all parties involved with implementing Annex II.

11 Implementing the revised Annex III: regulations for the prevention of pollution by harmful substances carried by sea in packaged forms

11.1 Brief explanation of the revised Annex III

The original Annex III entered into force on 1 July 1992. During the Earth Summit on Sustainable Development (ESSD, Rio 1992) it was decided that a globally harmonized system (GHS) for supply and use of all dangerous products, including chemicals should be developed. This development led to a major revision of both the Annex III and the International Maritime Dangerous Goods (IMDG) Code, regarding criteria for the identification of harmful substances in packaged form. The revised MARPOL Annex III was adopted by resolution MEPC.156(55), and entered into force on 1 January 2010.

Annex III applies to all ships to which MARPOL applies (see paragraph 3.3 of this manual) which carry harmful substances in packaged form. *Harmful substances* are those identified as marine pollutants in the IMDG Code or that meet the criteria in the appendix to Annex III. *Packaged form* means any form of containment other than the structure of the ship and includes packaging, freight containers, portable tanks and road and rail tank wagons as identified or specified in the IMDG Code.

Annex III prohibits the carriage of harmful substances except in accordance with the conditions laid down in the Annex. The conditions include requirements on packing, marking, labelling, documentation, and stowage and quantity limitations and exceptions relating to the safety of the ship or saving life at sea.

11.2 Implementation by means of the IMDG Code

Regulation 1 of the Annex makes reference to the IMDG Code for detailed requirements on the identification and carriage of harmful substances. The IMDG Code has been adapted so that it can be used as the practical means of implementation of Annex III. This is because the majority of harmful substances in packaged form (known as "marine pollutants") were already classified as dangerous goods in the IMDG Code.

For individual substances or materials, the IMDG Code consists of general requirements for each class of substances. Dangerous goods that are also marine pollutants, will have this fact clearly stated under the relevant column in the Dangerous Goods List. When a substance or material possess properties that may meet the criteria of a marine pollutant but is not included in the Dangerous Goods List, that substance or material shall, on the basis of self-classification, be transported as a marine pollutant in accordance with IMDG Code.

In cases involving marine pollutants, shippers will have to declare their shipment as a "marine pollutant" and comply with the Code's requirements. This will usually mean adding a special "marine pollutant" mark to the package.

Figure 7 – *Packages containing marine pollutants shall bear this mark as from 1 January 2012*

The marine pollutant mark (fish and tree), should be black on white or on a suitable contrasting background.

If a marine pollutant is not also classified as dangerous goods, it is listed in Class 9 of the IMDG Code. Such marine pollutant will need to be declared under the correct technical name. The packaging will have to conform to the requirements of the IMDG Code and be marked with the correct technical name, the UN number and the marine pollutant mark.

The method of marking and labelling shall be such that this information will still be identifiable on packages surviving at least three months immersion in the sea.

Full details of these requirements will be found in the latest edition of the IMDG Code. In order to facilitate an understanding of the foregoing, reference is made to chapter 2.10 of the IMDG Code. An extract of the relevant information is given in appendices 13, 14 and 15 of this manual.

On 1 October 2010, MEPC61 adopted further amendments to the revised Annex III, with an entry into force date of 1 January 2014. By MEPC.193(61) new, expanded criteria for marine pollutants, in line with Globally Harmonized System (GHS) criteria were adopted. In May 2012 the Maritime Safety Committee, at its ninetieth session, adopted amendments to the marine environment provisions, entering info force through amendment 36-12 to the IMDG Code. Resolution MSC.328(90) also has an entry into force date of 1 January 2014, however, it was agreed that contracting Governments to the Convention may apply the aforementioned amendments in whole or in part on a voluntary basis as from 1 January 2013.

11.3   Summary of actions and requirements for implementation of Annex III

11.3.1   Producers and shippers should ensure that, before any packaged harmful substance (marine pollutant) is offered for shipment, it is properly identified, packed, marked, labelled and documented in accordance with the IMDG Code.

11.3.2   Shipowners or operators should be satisfied that the packaging, marking, labelling and documentation are in order and should stow and secure the packages in accordance with Annex III and the IMDG Code. A special list, manifest or stowage plan, showing the location of the packages on board, should be kept on the ship with copies retained on shore in accordance with regulation 4.3.

11.3.3   No survey or certification of the ship is required under Annex III of MARPOL (although it is, under SOLAS, for dangerous goods). The detailed requirements referenced under regulation 1.3 must, however, be issued, which, in practice, means legislation must be in place so as to provide compliance with the IMDG Code (see paragraph 6.5.3). The marine administration should therefore inspect ships, in order to ensure this compliance, as a flag State duty and as a port State duty.

11.3.4   For regulation 8, port State control on operational requirements, reference is made to paragraph 21.1.4 of this manual.

12 Implementing Annex IV: regulations for the prevention of pollution by sewage from ships

12.1 Brief explanation of Annex IV

Annex IV entered into force on 27 September 2003. Before the Annex entered into force, it was recognized that major improvements needed to be made so that a sufficient number of States would ratify it. A revised Annex IV was approved by agreement among Member States. The revised Annex IV was adopted on 1 April 2004, by resolution MEPC.115(51), the first opportunity after the entry into force and entered into force on 1 August 2005.

The revised MARPOL Annex IV contained four chapters and one appendix. On 24 March 2006 the MEPC, by resolution MEPC.143(54), adopted an additional chapter on port State control consisting of a new regulation 13 on PSC on operational requirements. This amendment entered into force on 1 August 2007. By resolution MEPC.164(56) an amendment to regulation 11 on the discharge of sewage was adopted. This amendment expanded regulation 11 by inclusion of a reference to sewage originating from spaces containing living animals, and entered into force on 1 December 2008.

In 2010, it was proposed to include the principle of a special area also in Annex IV. By resolution MEPC.200(62), amendments to regulations 1, 9 and 11 were adopted, and a new regulation on reception facilities for passenger ships in special areas was inserted as regulation 12*bis*. These amendments have an entry into force date of 1 January 2013. Special area provisions are now incorporated in Annex IV and the Baltic Sea was designated as a special area, where the adoption of special mandatory methods for the prevention of pollution by sewage from ships is required. In this regard, new regulations on the discharge of sewage for passenger ships while in a special area were adopted. The date on which the requirements in respect of a special area will take effect depends on sufficient notifications to IMO from the Parties bordering the Baltic, on the availability of reception facilities for sewage. The concern raised was that there might be a possible lack of adequate and cost-effective technical onboard equipment to make it possible to meet the discharge standards for special areas. By resolution MEPC.218(63), MEPC made a call for the development of such equipment without delay.

The layout of Annex IV is currently as follows:

Chapter 1	General	Regulations 1 – 3
Chapter 2	Surveys and Certification	Regulations 4 – 8
Chapter 3	Equipment and control of discharge	Regulations 9 – 11
Chapter 4	Reception Facilities	Regulations 12 – 12bis
Chapter 5	Port State control	Regulation 13
Appendix	Form of International Sewage Pollution Prevention Certificate	

12.2 Chapter 1 – General

12.2.1 Definitions and application

Regulation 1 presents the definitions to be applied to this Annex. Regulation 2 describes to which ships the regulations of MARPOL Annex IV apply. The revised Annex applies to new ships (regulation 1.1 of Annex IV) engaged in international voyages, of 400 gross tonnage and above, or to new ships of less than 400 gross tonnage which are certified to carry more than 15 persons. Existing ships were required to comply with the provisions of the revised Annex IV five years after the date of entry into force of Annex IV. Annex IV entered into force on 27 September 2003, which means that as from 27 September 2008 there is no difference anymore between new and existing ships under Annex IV.

12.2.2 Exceptions

Exceptions to the discharge requirements are provided for in regulation 3. This is mainly related to exceptional circumstances like in case to secure the safety of a ship or saving life at sea or in case of combating specific pollution incidents.

12.3 Chapter 2 – Surveys and Certification

12.3.1 Surveys

In general, every ship that needs to comply with this Annex is subject to an initial survey and a renewal survey. An additional survey shall be made whenever any important repairs or renewals are made. Surveys are required to cover all applicable requirements of Annex IV (regulation 4), and the condition of the ship and its equipment are to be maintained to conform to the provisions of MARPOL. After a survey has been completed, no change shall be made in the structure, equipment, systems, fittings, arrangements or materials covered by the survey, without the sanction of the marine administration, except for the direct replacement of such equipment and fittings.

Appropriate measures (also likely to be surveys) to ensure compliance with the marine administration's requirements, must be established for ships for which Annex IV does not apply (regulation 4.2).

12.3.2 Issue or endorsement of certificate, issue or endorsement of certificate by another Government, form of certificate, duration and validity of certificate

After an initial or renewal survey an International Sewage Pollution Prevention (ISPP) Certificate shall be issued either by the Administration or by any persons or organization duly authorized by it. In any case the Administration assumes full responsibility for the certificate (regulation 5.2). An ISPP Certificate is required for ships trading internationally. A certificate is not required for ships in domestic trade but may be required by the marine administration in conjunction with the required surveys.

At the request of the Administration, a Government of a Party to MARPOL may survey a ship and, when satisfied with the results, may issue the ISPP Certificate. No ISPP Certificate shall be issued to a ship that flies the flag of a State which is not a Party to MARPOL (regulation 6.4). The ISPP Certificate shall be in the form as given in the appendix to Annex IV and shall be at least in English, French or Spanish. If the official language of the Administration is also used this shall prevail in case of a dispute or discrepancy (regulation 7).

In general an ISPP Certificate is valid for five years with a certain flexibility as described in regulation 8. A certificate will cease to be valid if relevant surveys are not completed within the period described in regulation 4 or upon transfer of the ship to the flag of another State.

12.4 Chapter 3 – Equipment and control of discharges

12.4.1 Sewage systems and standard discharge connections

The Annex requires that ships to which the Annex applies be equipped with either a sewage treatment plant or a sewage comminuting and disinfecting system or a sewage holding tank (regulation 9). The sewage treatment plant shall be of a type approved by the Administration, taking into account the standards and test methods developed by the Organization.

The MEPC, at its 55th session in October 2006, adopted revised Guidelines on implementation of effluent standards and performance tests for sewage treatment plants by resolution MEPC.159(55). The resolution invites Governements to implement the revised guidelines so that all equipment installed on board on or after

1 January 2010 meets the guidelines in so far as is reasonable and practicable and supersedes the Recommendation on international effluent standards and guidelines for performance tests for sewage treatment plants adopted by resolution MEPC.2(VI) in 1976. In light of the additional requirements under Annex IV regarding passenger ships in special areas, the standards as adopted by MEPC 55 were in need of updating. At MEPC 64, in October 2012, the 2012 Guidelines on implementation of effluent standards and performance tests for sewage treatment plants were adopted by resolution MEPC.227(64). However, before the adoption of the 2012 guidelines, concerns were raised whether equipment would be timely available. It was decided that a review on the availability would take place during MEPC 67 (second part of the year 2014).

Regulation 10 provides information on standards for discharge connections for both the ship and the reception facility.

12.4.2 Discharge of sewage

The discharge of sewage into the sea will be prohibited, except when the ship has in operation an approved sewage treatment plant, or is discharging comminuted and disinfected sewage using an approved system at a distance of more than three nautical miles from the nearest land, or is discharging sewage which is not comminuted or disinfected at a distance of more than 12 nautical miles from the nearest land. In October 2006, the MEPC, via resolution MEPC.157(55), adopted a recommendation on the standard for the maximum rate of discharge of untreated sewage from holding tanks when a ship is at a distance equal or greater than 12 nautical miles from the nearest land. Further, the discharge requirements for Annex IV were amended by resolution MEPC.164(56) to include sewage originating from spaces containing living animals. These amendments entered into force on 1 December 2008.

The discharge conditions outside a special area are given in tabular form below, which summarizes regulation 11A of Annex IV.

Table 6 – *Discharge requirements under Annex IV*

Sea area	Discharge criteria outside Special Area
Within 3 nautical miles from land	**No discharge** except from an approved sewage treatment plant certified to meet regulations 9.1.1 and 11A.1.2
Between 3 and 12 nautical miles from the nearest land	**No discharge** except either: (1) from an approved sewage treatment plant certified to meet regulations 9.1.1 and 11A.1.2; or (2) from an approved system for comminuting and disinfecting sewage meeting regulations 9.1.2 and 11A.1.1
More than 12 nautical miles from land	**Discharge** from either: (1) or (2) above; or sewage which is not comminuted or disinfected. Sewage that had been stored in holding tanks, or sewage originating from spaces containing living animals, shall not be discharged instantaneously but at a moderate rate when the ship is *en route* proceeding at not less than 4 knots and the rate of discharge is approved by the Administration. Reference is made to resolution MEPC.157(55), Recommendation on standards for the rate of discharge of untreated sewage from ships.

12.4.3 Discharge of sewage from passenger ships within a special area

As indicated in paragraph 12.1 of this manual, on 15 July 2011 the MEPC adopted amendments related to special area provisions and the designation of the Baltic Sea as a special area under MARPOL Annex IV. The consequential amendments to regulation 11 are reflected in a new regulation 11B and require that the discharge of sewage from passenger ships within a special area shall be generally prohibited:

   .1   for new passenger ships on, or after 1 January 2016; and

   .2   for existing passenger ships on, or after 1 January 2018.

An exception to the prohibition of discharge of sewage from passenger ships within a special area only applies when the ship has in operation an approved sewage treatment plant which shall be of a type approved by the Administration, taking into account the standards and test methods stipulated in the 2012 Guidelines on implementation of effluent standards and performance tests for sewage treatment plants. Also, the effluent shall not produce visible floating solids nor cause discoloration of the surrounding water.

A new passenger ship is a passenger ship for which the building contract is placed, or in the absence of a building contract, the keel of which is laid or which is in a similar stage of construction, on or after 1 January 2016 or, the delivery date of which is two years or more after 1 January 2016.

12.5   Chapter 4 – Reception facilities

Regulation 12 of Annex IV states that the Government of each Party to MARPOL undertakes to ensure the provision of adequate reception facilities for sewage at ports and terminals. The method by which a Government ensures this provision is covered in paragraph 6.5.8 and chapter 15 of this manual.

Chapter 4 is expanded by a regulation 12*bis* as a consequence of the introduction of special area provisions in Annex IV. Regulation 12*bis* deals with reception facilities for passenger ships in special areas.

12.6   Chapter 5 – Port State control

For regulation 13, port State control on operational requirements, reference is made to paragraph 21.1.4 of this manual.

12.7   Actions by the marine administration

The marine administration is required to take the following actions:

.1   carry out initial and renewal surveys or additional surveys as may be necessary (regulation 4);

.2   approve sewage treatment plants (regulation 9);

.3   approve ship sewage arrangements (regulation 9);

.4   establish appropriate measures for ships not engaged in international voyages (regulation 4.2);

.5   monitor the provision of adequate facilities at ports and terminals for the reception of sewage (regulation 12);

.6   issue certificates in the prescribed form (regulations 5, 6, 7 and 8); and

.7   notify the International Maritime Organization of all cases where the facilities provided were alleged to be inadequate.

12.8   Actions by the ports

Regulation 12 of Annex IV states that the Government of each Party to MARPOL undertakes to ensure the provision of adequate reception facilities for sewage (see also chapter 15). Regulation 12*bis* specifically addresses States bordering the Baltic Sea area.

12.9   Guidance

IMO has published guidelines on effluent standards and performance tests for sewage treatment plants (see paragraph 25.5.1) which should be referred to in connection with approval of such plants.

12.10   Summary of actions and requirements for implementation of Annex IV

12.10.1   Shipowners or operators should ensure that:

.1   the ship is equipped with a suitable sewage treatment plant or a comminuting and disinfecting system, or holding tanks, and a standard shore connection;

.2   the ship is surveyed and tests are carried out in line with the requirements of Annex IV;

.3   the ship is provided with an ISPP Certificate or document as required by the marine administration;

.4   the crew are trained to operate in accordance with the discharge criteria.

12.10.2   Marine administrations should:

.1   approve equipment and arrangements;

.2   carry out surveys;

.3   issue certificates; and

.4   establish measures as required by the Annex for ships not engaged in international voyages.

12.10.3   Ports should:

.1   assess requirements for sewage reception facilities; and

.2   ensure that adequate reception facilities are available.

13 Implementing the revised Annex V: regulations for the prevention of pollution by garbage from ships

13.1 Brief explanation of Annex V

Annex V applies to all ships, which means all vessels of any type whatsoever operating in the marine environment, from merchant ships to fixed and floating platforms to non-commercial ships like pleasure crafts and yachts. Under the regulations of the original Annex V the discharge of garbage into the sea is prohibited except for certain types of garbage. In this respect, a rough distinction could be made between plastics, materials that float and other garbage. For example, plastics were not allowed to be discharged into the sea but for materials that float and for other garbage, no quantitative limitations existed.

On 15 July 2011, by resolution MEPC.201(62), a revised Annex V was adopted. This revised Annex V has an entry into force date of 1 January 2013. In this chapter the focus will be on the revised Annex V which contains ten regulations.

The main differences between the original and the revised Annex V are that the latter introduced the principle of a general discharge prohibition with very limited exceptions, that it has clear definitions, and that cargo residues of solid bulk cargoes are better defined.

As a consequence of the revision of Annex V, also the guidelines to the original Annex V needed to be amended. Revised guidelines were adopted on 2 March 2012 by resolution MEPC.219(63) under the title: 2012 Guidelines for the implementation of MARPOL Annex V.

An important item in the guidelines is the management of cargo residues of solid bulk cargoes, paragraph 3 of the 2012 guidelines. Cargo residues are considered harmful to the marine environment and therefore subject to regulations 4.1.3 and 6.1.2.1 of the revised Annex V, if they meet the parameters in paragraph 3.2 of the 2012 guidelines. In this respect a link has been made with the International Maritime Solid Bulk Cargoes Code (IMSBC Code).

13.2 Regulation by regulation

13.2.1 Regulation 1 – Definitions

Regulation 1 presents the definitions to be applied to this Annex. One of the goals of the revision was that all expressions used in the Annex should be covered by a definition. This goal has been achieved which makes the revised Annex transparent and easy to implement and enforce.

13.2.2 Regulation 2 – Application

As indicated under 13.1, the provisions of the Annex apply to all ships.

13.2.3 Regulation 3 – General prohibition on discharge of garbage into the sea

The principle of the revised Annex V is that the discharge of all garbage into the sea is prohibited, except when regulation 7, exceptions, is applicable. In addition, specifically identified types of garbage, like food wastes, are allowed to be discharged under strict conditions.

13.2.4 Regulation 4 – Discharge of garbage outside special areas

The discharge of garbage into the sea outside special areas is prohibited, however, there are provisions which allow the discharge of individually identified types of garbage like food waste, cargo residues and animal carcasses. Next to specific discharge requirements for these types of garbage, discharges are only permitted when the ship is *en route* and as far as practicable from the nearest land. To discharge cargo hold washwater in this situation, the ship must be *en route* and the discharge must take place not less than 12 nautical miles from the nearest land. Cleaning agents and additives contained in cargo hold, deck, and external surface washwater may be discharged into the sea outside special areas, but only if these substances are not harmful to the marine environment in accordance with paragraph 1.7.5 of the 2012 *Guidelines for the implementation of MARPOL Annex V.*[*] The discharge requirements also apply when the "port of departure" and the "next port of destination" are the same port.

13.2.5 Regulation 5 – Special requirements for discharge of garbage form fixed or floating platforms

Additional restrictions apply to fixed or floating platforms while they are engaged in exploration or exploitation of the sea-bed, and to other ships within 500 m of such platforms.

13.2.6 Regulation 6 – Discharge of garbage within special areas

There are at present eight special areas (SAs) under Annex V: the Mediterranean Sea, the Black Sea area, the Baltic Sea area, the Red Sea area, the Gulfs area, the North Sea area, the Antarctic area and the Wider Caribbean Region.

Discharges of food waste and certain types of cleaning agents or additives and cargo residues not including any substances classified as harmful to the marine environment are allowed, however, strict conditions apply.

These prohibitions and restrictions are shown in tabular form in table 7.

_____
[*] See IMO publication, sales number IB656E.

Table 7 – *Summary of restrictions to the discharge of garbage into the sea under regulations 4, 5 and 6 of MARPOL Annex V*

Garbage type[1]	All ships except platforms[4]		Offshore platforms located more than 12 nautical miles from nearest land and ships when alongside or within 500 m of such platforms[4] Regulation 5
	Outside Special Areas Regulation 4 (Distances are from the nearest land)	Within Special Areas Regulation 6 (Distances are from nearest land or nearest ice-shelf)	
Food waste comminuted or ground[2]	≥ 3 nautical miles, *en route* and as far as practicable	≥ 12 nautical miles, *en route* and as far as practicable[3]	**Discharge permitted**
Food waste not comminuted or ground	≥ 12 nautical miles, *en route* and as far as practicable	Discharge prohibited	**Discharge prohibited**
Cargo residues[5, 6] not contained in washwater	≥ 12 nautical miles, *en route* and as far as practicable	Discharge prohibited	**Discharge prohibited**
Cargo residues[5, 6] contained in washwater		≥ 12 nautical miles, *en route* and as far as practicable (subject to conditions in regulation 6.1.2)	
Cleaning agents and additives[6] contained in cargo hold washwater	Discharge permitted	≥ 12 nautical miles, *en route* and as far as practicable (subject to conditions in regulation 6.1.2)	**Discharge prohibited**
Cleaning agents and additives[6] in deck and external surfaces washwater		Discharge permitted	
Animal carcasses (should be split or otherwise treated to ensure the carcasses will sink immediately)	Must be *en route* and as far from the nearest land as possible. Should be > 100 nautical miles and maximum water depth	**Discharge prohibited**	**Discharge prohibited**
All other garbage including plastics, synthetic ropes, fishing gear, plastic garbage bags, incinerator ashes, clinkers, cooking oil, floating dunnage, lining and packing materials, paper, rags, glass, metal, bottles, crockery and similar refuse	Discharge prohibited	**Discharge prohibited**	**Discharge prohibited**

[1] When garbage is mixed with or contaminated by other harmful substances prohibited from discharge or having different discharge requirements, the more stringent requirements shall apply.

[2] Comminuted or ground food wastes must be able to pass through a screen with mesh no larger than 25 mm.

[3] The discharge of introduced avian products in the Antarctic area is not permitted unless incinerated, autoclaved or otherwise treated to be made sterile.

[4] Offshore platforms located 12 nautical miles from nearest land and associated ships include all fixed or floating platforms engaged in exploration or exploitation or associated processing of seabed mineral resources, and all ships alongside or within 500 m of such platforms.

[5] Cargo residues means only those cargo residues that cannot be recovered using commonly available methods for unloading.

[6] These substances must not be harmful to the marine environment.

13.2.7 Regulation 7 – Exceptions

The general exception of the discharge requirements necessary for the purpose of securing the safety of a ship and those on board or saving life at sea is also applicable for garbage.

In addition, there is an exception to the *en-route* requirement for the discharge of food waste, where it is clear that retention on board of this waste presents an imminent health risk to the people on board.

13.2.8 Regulation 8 – Reception facilities

This regulation states that each Party undertakes to ensure that provision of adequate facilities at ports and terminals for the reception of garbage without causing undue delay to ships, and according to the needs of ships using them. The method by which a Government ensures this provision is covered in paragraph 6.5.8 and chapter 15 of this manual.

13.2.9 Regulation 9 – Port State control

This regulation pertains to port State control on operational requirements. Reference is made to paragraph 21.1.4 of this manual.

13.2.10 Regulation 10 – Placards, garbage management plans and garbage record-keeping

Unlike those of the other Annexes of MARPOL, the provisions of Annex V address discharges from ships and do not stipulate equipment requirements. The revised Annex V (regulation 10.1) requires every ship of 12 m or more in length overall and fixed and floating platforms to display placards indicating the discharge requirements under the revised Annex V. Regulation 10.2 requires every ship of 100 gross tonnage and above, and every ship which is certified to carry 15 persons or more, and fixed or floating platforms to have in place a garbage management plan that the crew shall follow The plan shall be developed in accordance with the guidelines adopted by resolution MEPC.220(63): 2012 Guidelines for the development of garbage management plans and shall be written in the working language of the crew. The plan must designate the person in charge of its execution and contain written procedures for minimizing, collecting, storing, processing and disposing of garbage, including the use of the equipment on board.

Regulation 10.3 requires every ship of 400 gross tonnage and above and every ship which is certified to carry 15 persons or more engaged in voyages to ports or offshore terminals under the jurisdiction of another Party and every fixed and floating platform to be provided with a Garbage Record Book (GRB). Each discharge operation or completed incineration on board must be duly recorded.

13.2.11 Appendix to the revised Annex V

The form of the Garbage Record Book is reflected in the appendix to the revised Annex V. In paragraph 4 of that appendix the entries which shall be made in the GRB are provided.

13.3   Actions and requirements for implementation of Annex V

13.3.1   Shipowners and operators should ensure that:

.1   arrangements on the ship are adequate for dealing with garbage:

For all ships it is essential to allocate a space on board to store garbage until it can be disposed of at reception facilities, or until it can be incinerated. The provision of such storage spaces, processing equipment and means of disposal is the shipowner's or operator's responsibility. Such arrangements will vary from minimal in small ships, to properly designed spaces, processing plants or incinerators and organized handling on large passenger ships;

.2   the crew are properly instructed and trained to comply with the discharge conditions;

.3   every ship of 400 gross tonnage and above, and every ship certified to carry 15 persons or more, has in place a garbage management plan that the crew shall follow, developed in accordance with IMO guidelines and written in the working language of the crew. The plan must designate the person in charge of its execution and contain written procedures for minimizing, collecting, storing, processing and disposing of garbage and for the use of the equipment on board;

.4   every ship which is required to have a Garbage Record Book (GRB) which is maintained in accordance with regulations under the Annex. Each discharge operation or completed incineration must be duly recorded;

.5   every ship of 12 m or more in length overall displays placards to notify the crew and passengers of the discharge requirements under Annex V.

13.3.2 The marine administration should:

.1   provide advice to its own-flag ships;

.2   examine on board arrangements during inspections;

.3   monitor the provision of adequate facilities at ports and terminals for the reception of garbage;

.4   notify the International Maritime Organization of all cases where the facilities provided were alleged to be inadequate;

.5   investigate infringements; and

.6   prosecute offenders.

13.3.3 Ports should:

.1   assess requirements for reception facilities for garbage; and

.2   ensure that adequate reception facilities are available.

Regulation 8 states that the Government of each Party to Annex V undertakes to ensure the provision of adequate reception facilities for garbage from ships using its ports and terminals. For provisions of reception facilities, reference is made to chapter 15 of this manual.

14 Implementing Annex VI: regulations for the prevention of air pollution from ships

14.1 Brief explanation of Annex VI

Annex VI applies to all ships, except where expressly provided otherwise in several regulations. In contrast to the other MARPOL Annexes, Annex VI controls a range of different pollutant streams together with certain aspects related to ship operation which can themselves result in air pollution. Air pollution does not have the direct cause and effect associated with, for example, an oil spill incident. Rather, it is the cumulative effect from shipping in general which contributes to the overall air quality encountered by populations at large and which affects both the natural and built environments potentially directly affecting areas a considerable distance from the point of discharge and therefore remote from the sea.

The controls within Annex VI cover:

- ozone-depleting substances released from refrigeration and fire-fighting systems and equipment. Such substances are also contained in some types of insulation foams;

- nitrogen oxides from diesel engine combustion;

- sulphur oxides and particulate matter emissions from the combustion of fuel oils which contain sulphur;

- volatile organic compounds, the hydrocarbon vapours displaced from tanker cargo spaces;

- shipboard incineration;

- fuel oil quality in so far as it relates to a number of air quality issues; and

- energy efficiency for ships.

Compliance with the relevant requirements of Annex VI is indicated by the issue of an International Air Pollution Prevention (IAPP) Certificate for ships of 400 gross tonnage and above and for platforms and drilling rigs engaged in international voyages. Ships of 400 gross tonnage and above are required to be issued with an International Energy Efficiency (IEE) Certificate. For ships of less than 400 gross tonnage appropriate measures may be developed by an Administration in order to demonstrate the necessary compliance.

14.2 Air pollution and energy efficiency control and related survey and certification requirements

An International Air Pollution Prevention (IAPP) Certificate is required for all ships of 400 gross tonnage and above for which the flag State is a Party to Annex VI. In the case of ships constructed on or after the date of entry into force of Annex VI, this is to be issued, following satisfactory completion of the initial survey and prior to entry into service. In the case of existing ships, the initial survey leading to the issue of an IAPP Certificate is to be undertaken no later than the first scheduled dry-docking after the date of entry into force of the Annex for a particular Party but in no case later than three years after that date. In respect of existing ships, the relevant equipment and operational requirements apply whether or not the IAPP Certificate has been issued. Thereafter, annual/intermediate and renewal surveys are scheduled in accordance with the harmonized system of survey and certification together with any additional surveys which may be required following repairs, replacements or the installation of additional equipment.

Specific guidelines for surveys related to Annex VI under the survey guidelines under the Harmonized System of Survey and Certification, are included in the consolidated version of these survey Guidelines (see Assembly resolution 1053(27)). Also specific guidelines for port State control exist (MEPC.181(59)).

On 15 July 2011, amendments to Annex VI were adopted by resolution MEPC.203(62). These amendments provide regulations on energy efficiency for ships under MARPOL Annex VI, mainly via a new chapter 4, and have an entry into force date of 1 January 2013 (paragraph 14.2.7). The inclusion of the new chapter 4 has also led to consequential amendments to several other chapters such as chapter 2, Survey, Certification and Means of Control. In addition to the IAPP Certificate, every ship to which the new chapter 4 applies, shall have an International Energy Efficiency (IEE) Certificate. Prior to issuing the IEE Certificate additional surveys are required. Major difference between the two certificates is that the IEE Certificate is valid throughout the life of the ship. It only ceases to be valid if the ship is withdrawn from its service or if a new certificate is issued following a major conversion and upon transfer of the ship to the flag of another State.

The new chapter 4 mainly applies to new ships with the exception of the requirement of the Ship Energy Efficiency Management Plan (SEEMP). The verification of the requirement to have a SEEMP is also applicable to existing ships, and shall take place at the first intermediate or renewal survey, whichever is the first, on or after 1 January 2013.

The following sections provide overviews of the controls applicable to each of the respective controlled pollution streams covered within chapter 3 of Annex VI and energy efficiency measures covered within chapter 4.

14.2.1 Ozone-depleting substances (ODSs) – regulation 12

Included within the definition of ODSs are the chlorofluorocarbons (CFCs) and halons used respectively in older refrigeration and fire-fighting systems and portable equipment. ODSs were also used as the blowing agent in some insulation foams. Hydrochlorofluorocarbons (HCFCs) were introduced as an intermediate replacement for CFCs but are themselves still classed as ODSs. As part of a worldwide movement, the production and use of all these materials is being phased out under the provisions of the Montreal Protocol.

The controls in this regulation do not apply to sealed equipment without refrigerant charging connections or removable components; this regulation typically covers items such as small, domestic-type, refrigerators, air conditioners and water coolers.

No CFCs or halon-containing system or equipment is permitted to be installed on ships constructed on or after 19 May 2005, and no new installation of the same is permitted on or after that date on existing ships. Similarly, no HCFC-containing system or equipment is permitted to be installed on ships constructed on or after 1 January 2020, and no new installation of the same is permitted on or after that date on existing ships.

Existing systems and equipment are permitted to continue in service and may be recharged as necessary. However, the deliberate discharge of ODSs to the atmosphere is prohibited. When servicing or decommissioning systems or equipment containing ODSs the gases are to be duly collected in a controlled manner and, if not to be reused on board, are to be landed to appropriate reception facilities for banking or destruction. Any redundant equipment or material containing ODSs is to be landed ashore for appropriate decommissioning or disposal. The latter also applies when a ship is dismantled at the end of its service life.

Additionally, for ships with systems or equipment containing ODSs and which are required to have an IAPP Certificate, an ODSs Record Book is to be maintained in which is recorded any related supply, recharging, repair, discharge operations.

14.2.2 Nitrogen oxides (NO_x) – regulation 13

The control of diesel engine NO_x emissions is achieved through the survey (regulation 5.3.2) and certification (regulation 13.8) requirements leading to the issue of an Engine International Air Pollution Prevention (EIAPP) Certificate and the subsequent demonstration of in-service compliance in accordance with the requirements of the mandatory regulations 13.8 and 5.3.2 respectively, NO_x Technical Code 2008 (resolution MEPC.177(58)).

The NO_x control requirements of Annex VI apply to installed marine diesel engines of over 130 kW output power other than those used solely for emergency purposes, irrespective of the tonnage of the ship onto which such engines are installed. Definitions of "installed" and "marine diesel engine" are given in regulations 2.12 and 2.14 respectively. Different levels (Tiers) of control apply based on the ship construction date, defined in regulations 2.19 and hence 2.2. Within any particular Tier the limit value is determined based on the engine's rated speed:

Table 8 – *Standards for emissions of NO_x by marine diesel engines, by Tier*

Tier	Ship construction date (on or after)	total weighted cycle emission limit (g/kWh) n = engine's rated speed (rpm)		
		$n < 130$	$n = 130 - 1,999$	$n \geq 2,000$
I	1 January 2000	17.0	$45 \cdot n^{-0.2}$ e.g., 720 rpm – 12.1	9.8
II	1 January 2011	14.4	$44 \cdot n^{-0.23}$ e.g., 720 rpm – 9.7	7.7
III	1 January 2016[*]	3.4	$9 \cdot n^{-0.2}$ e.g., 720 rpm –2.4	2.0

[*] This date is subject to a technical review to be concluded in 2013 (regulation 13.10).

The Tier III controls apply only to the specified ships while they operate in Emission Control Areas (ECAs) established to limit NO_x emissions. Outside such areas, the Tier II controls apply. At the time of developing this manual the North American ECA and the United States Caribbean Sea ECA (see paragraph 3.27 of this manual), have been established. In accordance with regulation 13.5.2, certain small ships would not be required to install Tier III engines. It should be noted that ships constructed on or after 1 January 2016 shall comply with the NO_x emission limits specified in regulation 13.5.1.1, when operating within the North American ECA or within the United States Caribbean Sea ECA.

The emission value for a diesel engine is to be determined in accordance with the NO_x Technical Code 2008, in the case of Tier II and Tier III limits. Most Tier I engines have been certified to the earlier, 1997, version of the NO_x Technical Code which, in accordance with MEPC.1/Circ.679, was allowed to be used in certain cases until 1 January 2011. Certification issued in accordance with the 1997 NO_x Technical Code would still remain valid over the service life of such engines.

As per the 2008 NO_x Technical Code, the engine family or engine group approval concepts may be adopted to avoid certification testing of every engine for compliance with the limits of regulation 13. An engine may be certified on an individual, engine family or engine group basis in accordance with one or more of the four duty test cycles as given in appendix II of the Annex. In the case of application of the engine family or engine group concept to engines, it is the parent engine which is actually emissions tested. The parent engine is the engine which has the combination of rating (power and speed) and NO_x critical components, settings and operating values that results in the highest NO_x emission value or, where more than one test cycle is to be certified, values, which, to be acceptable, must be no higher than the applicable Tier limit value. Subsequent series engines, member engines, are thereafter constructed with a rating, components, settings and operating values within the bounds established for the respective engine family or engine group. Generally all new engine certification leading to the issue of an EIAPP Certificate is undertaken at the engine builder's works where the necessary pre-certification survey takes place.

Subsequently, a diesel engine having an EIAPP Certificate is approved by, or on behalf of (almost all engine certification work is delegated to Recognized Organizations – see chapter 23), the flag State of the ship onto which it is to be installed, to a stated Tier for one or more duty test cycles, for a particular rating or rating range, and with defined NO_x critical components, settings and operating values, including options, if applicable. Any amendments to these aspects are to be duly approved and documented.

For each NO_x certified diesel engine there must be on board an approved technical file (NO_x Technical Code 2008, 2.3.4), which both defines the engine as approved and provides the applicable survey regime together with any relevant approved amendment documentation. At the time of development of this manual virtually all engines were surveyed using the parameter check method (NO$_x$ Technical Code 2008, 2.4.3.1), whereby the actual duty, rating and NO$_x$ critical components, settings and operating values are checked against the given data in the technical file. A key document in the parameter check procedure is the Record Book of Engine Parameters (NO$_x$ Technical Code 2008, 6.2.2.8), which is maintained to record all replacements and changes to NO$_x$ critical components, settings and operating values. Engine surveys are undertaken on completion of manufacture and subsequently as part of the overall ship survey process. NO$_x$ Technical Code 2008, appendix II, presents flowcharts illustrating the aspects checked at the various survey stages.

Case in which a diesel engine has been subjected to "major conversion", as defined in regulation 13.21, are addressed by regulations 13.2.1 – 13.2.3. Three classes of major conversion are given, of which "substantial modification" (NO$_x$ Technical Code 2008, 1.3.2) and up-rating (when the maximum continuous rating of the original certification of the engine is increased by more than 10% compared to the maximum continuous rating of the original certification of the engine) involve changes to an existing installed engine. Under these two circumstances, the relevant Tier is that applicable on the construction date of the ship onto which the engine is installed, however, in the case of ships constructed before 1 January 2000, the Tier I standard applies. The other type of major conversion is the installation of a replacement or additional engine. When a marine diesel engine is replaced with a non-identical marine diesel engine or an additional marine diesel engine is installed, then the Tier appropriate to the date of installation applies. However, subject to acceptance by the Administration and taking into account guidelines to be developed by the Organization, in some circumstances it would be permitted to install a replacement engine at a Tier II standard, as opposed to one certified to Tier III (regulation 13.2.2). In the case of an identical replacement engine, the Tier appropriate to the ship construction date applies.

The revised Annex VI (regulations 13.7.1 – 13.7.5) introduced NO$_x$ certification in retrospect, in the case of diesel engines of more than 5,000 kW power output and a per cylinder displacement of 90 litres and above installed on ships constructed on or after 1 January 1990 and before 1 January 2000. This scheme will generally affect only the main engines on such ships; in current medium-speed engine designs, the 90 litre per cylinder criterion represents, engines with a bore of 460 mm and above. The certification relies on the existence of an Approved Method (NO$_x$ Technical Code 2008, 1.3.17), a method for a particular engine, or a range of engines that, when applied to the engine, will ensure that the engine complies with the existing NO$_x$ limit as detailed in regulations 13.7.1 – 13.7.5. Provided that a Party, not necessarily the ship's flag State, has certified an "Approved Method" for a marine diesel engine and has advised IMO of that certification, the engine concerned is required to have an emission value no higher than the relevant Tier I level (regulation 13.7.4). The Approved Method must be applied no later than the first renewal survey which occurs more than 12 months after deposition of the advice to IMO (regulation 13.7.2). If the shipowner of a ship on which the approved method is to be installed can demonstrate that the approved method is not commercially available, despite best efforts to obtain it, then it is to be installed no later than the next annual survey after which it has become commercially available. Regulation 13.7.5 places constraints on the approved method that limit its cost and the detrimental effects it might have on engine power and fuel consumption. Further requirements are given in chapter 7 of the NO$_x$ Technical Code 2008, including an outline of the approved method file, which must be retained with the engine.

Several certified approved methods have now been notified to the Organization and subsequently issued in circulars. Approved methods include such aspects as changing the engine's fuel injection nozzles. Shipowners (and also surveyors and port State control inspectors) will need to remain vigilant over the service life of engines that might be required to install an approved method, with regard to the availability of such arrangements and to ensure that they are duly fitted and thereafter retained, as required. For those engines where an approved method exists, regulation 13.7.1.2 provides an alternative option whereby the engine is instead certified in accordance with the conventional NO$_x$ Technical Code 2008 requirements.

14.2.3 Sulphur oxides (SO$_x$) and particulate matter – regulation 14

SO$_x$ and particulate matter emission controls apply to all fuel oil used on board, as defined in regulation 2.9 and include main and all auxiliary engines together with such items as boilers and inert gas generators. These controls are divided into those applicable inside Emission Control Areas (ECAs) established to limit the emission of SO$_x$ and particulate matter, and those controls applicable outside such areas. The targets are primarily achieved by limiting the maximum sulphur content of fuel oils as bunkered and subsequently used on board. Fuel oil sulphur limits (expressed in terms of % m/m – that is by weight) are subject to a series of step changes over the years. The schedules are provided in regulations 14.1 and 14.4 and are reflected in table 9 in a tabular form:

Table 9 – *Limits on sulphur content of any fuel used on board ships*

Outside an ECA established to limit SO$_x$ and particulate matter emissions	Inside an ECA established to limit SO$_x$ and particulate matter emissions
4.50% m/m prior to 1 January 2012	1.50% m/m prior to 1 July 2010
3.50% m/m on and after 1 January 2012	1.00% m/m on and after 1 July 2010
0.50% m/m on and after 1 January 2020[*]	0.10% m/m on and after 1 January 2015

[*] Depending on the outcome of a review, to be concluded in 2018, as to the availability of the required fuel oil, this date could be deferred to 1 January 2025.

At the time of development of this manual, the ECAs established to limit SO$_x$ and particulate matter emissions (see paragraph 3.27) are:

– Baltic Sea area – as defined in MARPOL Annex I

– North Sea area – as defined in MARPOL Annex V

– North American Emission Control Area – as defined in appendix VII of Annex VI

– United States Caribbean Sea Emission Control Area (with an entry into force date of 1 January 2013).

MEPC.1/Circ.723 provides further information on the North American Emission Control Area (ECA) under MARPOL Annex VI, and in particular related to the effective date of the requirements for SO$_x$ which was 1 August 2012. While ships are operating in the North American ECA, the sulphur content of fuel oil used on board ships shall not exceed 1.00% m/m on and after 1 August 2012, and 0.10% m/m on and after 1 January 2015.

In accordance with the provisions of regulation 14.7 of MARPOL Annex VI, the requirements within the United States Caribbean Sea ECA for SO$_x$ and particulate matter will be effective on 1 January 2014 (MEPC.1/Circ.755). While ships are operating in the United States Caribbean Sea ECA, the sulphur content of fuel oil used on board ships shall not exceed 1.00% m/m on and after 1 January 2014, and 0.10% m/m on and after 1 January 2015. For ships, built on or before 1 August 2011, that are powered by propulsion boilers that were not originally designed for continued operation on marine distillate fuel or natural gas, the above sulphur requirements may not be applied prior to 1 January 2020 (regulation 14.4.4 of MARPOL Annex VI).

Ships which operate both outside and inside these ECAs consequentially will likely operate on different fuel oils (equivalent compliance methods can also be used and may become more prevalent) in order to comply with the respective limits (see regulation 14.6). They are required to have on board written procedures as to how the change-over between fuel oils is to be carried out. A ship is required to have fully changed over to using ECA compliant fuel oil prior to entering the ECA from an area outside the ECA Similarly change-over from the ECA compliant fuel oil is not to commence until after the ship has exited the ECA. At each change-over, the quantities of the ECA compliant fuel oils on board shall be recorded, together with the date, time and position of the ship when either the change-over prior to entry into an ECA has been completed or the change-over after exit from an ECA has been commenced. These records are to be made in a logbook as prescribed by the ship's Administration. In the absence of any specific requirement in this regard, the record could be made in, for example, the ship's Oil Record Book mandatory under MARPOL Annex I.

The first level of control on fuel oil is on the actual sulphur content of the fuel oils as bunkered. This value is to be stated by the fuel oil supplier on the bunker delivery note. This value, together with other related aspects, is directly linked to the fuel oil quality requirements as covered under regulation 18 (see paragraph 14.2.6 below). In the case of ECA-compliant fuel oils, the ship's crew shall then ensure that these fuel oils not become mixed with other, higher-sulphur-content fuel oils. For example, it should be avoided to bunker ECA-compliant fuel oil into part filled bunker, settling or service tanks containing non-ECA-compliant fuel oil. Also in the course of transfer operations, ECA compliant fuel oils should not be mixed with non-ECA-compliant fuel oils, so that the fuel oil as actually used within an ECA does not exceed the applicable limit.

Regulation 14 provides the limit values of sulphur content fuel oil and required that rigorous records be kept and bunkers not be mixed, to ensure the limits not be breached. However, there are other means by which equivalent levels of SO_x and particulate matter emission control, both outside and inside ECAs, could be achieved. Equivalent methods may be divided into primary (formation of the pollutant is avoided) and secondary (the pollutant is formed and then removed, prior to the discharge of the exhaust gas stream to the atmosphere). Regulation 4.1 allows for the application of such methods, subject to approval by the Administration. In approving such equivalents, an Administration should take into account any relevant guidelines developed by the Organization pertaining to them (regulation 4.3). At the time of development of this manual there were no guidelines in respect of any primary methods (which could encompass, for example, dual fuel (gas/liquid) use). In terms of secondary control methods, guidelines have been adopted for exhaust gas cleaning systems which operate by water-washing the exhaust gas stream prior to discharge to the atmosphere (MEPC.184(59)). In using such arrangements, there would be no constraint on the sulphur content of the fuel oils as bunkered other than that identified in the system's certification.

14.2.4 Volatile organic compounds (VOCs) – regulation 15

This regulation applies only to tankers. There are two aspects of VOC control within this regulation. In the first, certain ports or terminals can control the emissions of VOCs to the atmosphere by utilizing a vapour emission control system (VECS), in accordance with regulations 15.1 – 15.5 and 15.7. Where such emissions are regulated, both the shipboard vapour emission collection system and shore vapour emission control arrangements should take into account the safety standards for such systems developed by the Organization (MSC/Circ.585, "Standards for vapour emission control systems"). A Party may choose to apply such controls only to particular ports or terminals under its jurisdiction and only to certain sizes of tankers or types of cargo. Where such controls are in place at particular ports or terminals, tankers not fitted with a vapour emission collection system may be accepted for a period of up to three years from the effective date of control, as described in regulation 15.2. Where a VECS is mandated, the relevant Party is to notify IMO of the requirement and its date of implementation. At the time of developing this manual some notifications had been made.

The second aspect of this regulation, regulation 15.6, requires that all tankers carrying crude oil have an approved and effectively implemented ship-specific VOC Management Plan covering at least the information given in the regulation. See resolution MEPC.185(59) for guidelines on the development of these plans, and circular MEPC.1/Circ.680 for technical information on systems and operation to assist in development of such arrangements.

14.2.5 Shipboard incineration – regulation 16

This regulation applies to shipboard incinerators only, except as provided in regulation 16.4. Regulations 16.1 – 16.4 cover on board incineration in general and hence are applicable to all ships, whereas regulations 16.6.1 – 16.9 are specific to incinerators installed on ships constructed on or after 1 January 2000 or to incinerators installed on existing ships on or after that date.

Regulation 16.1 requires that incineration only be undertaken in equipment designed for that purpose. Regulation 16.2 prohibits the incineration of certain listed materials and can therefore be seen as complementary to the Annex V requirements in respect of the processing of ship-generated garbage. The disposal of polyvinyl chlorides (PVCs) by incineration is restricted to units which are type approved to either MEPC.76(40) or MEPC.59(33) (regulation 16.3). Regulation 16.4 recognizes that, while incineration of ship-generated sewage sludge and sludge oil could alternatively be undertaken in the main or auxiliary power plant or boilers, in those cases it is not to be undertaken within ports, harbours or estuaries.

Regulation 16.6.1 generally requires that incinerators installed on ships constructed on or after 1 January 2000 or units which are installed on existing ships on or after that date be approved by the Administration in accordance with appendix IV to Annex VI taking into account resolution MEPC.76(40), Standard specification for shipboard incinerators, as modified by resolution MEPC.93(45). For these incinerators, operating manuals are to be maintained with the unit (regulation 16.7), and training in correct operation of the units in accordance with the manual is to be given (regulation 16.8). Regulation 16.9 requires that the incinerators only be operated when certain prescribed temperatures have been achieved.

14.2.6 Fuel oil quality – regulation 18

In general, this regulation is not directed to ships, but rather relates to fuel oil suppliers and their control by the appropriate authorities, together with other regulatory aspects. In particular, the requirements of regulations 18.1, 18.2.1 – 18.2.5, 18.4, 18.5, 18.8.2, 18.9 and 18.10, together with aspects of regulations 18.8.1, should be seen as supportive of regulation 14 in respect of those aspects which are outside the control of the shipowner.

For those ships that are required to have an IAPP Certificate, regulations 18.6 and 18.8.1 contain specific ship-related actions concerning, respectively, retaining bunker delivery notes on board for a period of not less than three years following delivery of the fuel oil, a regulation that may be relaxed under conditions specified in regulation 18.11, and retaining, under the ship's control (and therefore not necessarily on board), representative fuel oil samples until the subject fuel oil is substantially consumed, but for not less than 12 months from the date of delivery. These requirements apply whether or not regulation 14 is complied with by means of bunkering fuel oils which do not exceed the stated limits.

The 2009 guidelines for the sampling of fuel oil for determination of compliance with the revised MARPOL Annex VI were issued by resolution MEPC.182(59) to take into account the revision of the Annex. Paragraphs 8 and 9 of those guidelines include specific actions to be taken by the personnel of the ship. It should be noted that these guidelines are recommendatory to the regulating authority, and, as such, local legislation covering the control of fuel oil suppliers in respect of Annex VI-related issues may not directly follow all aspects as given in these guidelines. For example, in the case of fuel oil sampling location, the relevant authority may accept other equivalent arrangements which are duly controlled as required.

It is necessary that oversight by the personnel of the ship is possible on both the bunker delivery note and the representative fuel oil sample. In accordance with paragraphs 2.1.1.12 and 2.1.5 of the 2009 Guidelines for port State control under the revised MARPOL Annex VI (resolution MEPC.181(59)), instances when the bunker delivery note does not contain the information as required by appendix V of revised Annex VI or when the representative sample has not been drawn, labelled or sealed in accordance with the relevant guidelines are to be duly documented and advised to the ship's flag State Administration with copies to the bunkering port authorities and the bunker supplier, and with a further copy retained on board together with any relevant commercial documentation.

Another responsibility assumed by the ship is in cases when the bunkering port is located in a country not party to Annex VI. In such a situation, apart from commercial considerations, there is no direct requirement for such ports to comply with the various requirements of regulation 18. Hence, it is common for shipowners, when ordering bunkers, at a minimum to insert clauses to the effect that the fuel oil supply process is to be in accordance with the requirements of Annex VI and with a specified maximum sulphur content appropriate to the particular intended area of operation.

The other aspect of regulation 18 that places responsibilities on shipowners regards the availability of fuel oil. Regulations 18.2.1 – 18.2.5 provide for situations in which the required fuel oil is not available locally. In other words, there is only fuel oil which does not meet the required maximum sulphur limit as given in regulation 14. The shipowner must have made his or her best efforts to obtain the required fuel oil, and this effort should be taken into account by Parties when considering what action to take, or not to take, in the case of a ship using fuel oil the composition of which does not comply with the regulations.

Regulation 18.9 together with regulations 18.1, 18.3, 18.4 and the first part of 18.5 refer to the application of local controls on fuel oil suppliers, while regulations 18.7.1 – 18.7.2, 18.8.2 – and hence appendix VI – and 18.10 refer to the application of port State controls.

14.2.7 Regulations on energy efficiency for ships

On 15 July 2011, regulations on energy efficiency for ships were included in Annex VI by resolution MEPC.203(62) with an entry into force date of 1 January 2013. The additional regulations are reflected in a new chapter 4, however, it should be noted that the inclusion of these additional regulations also have an effect on regulations 1, 2, 5, 6, 7, 8, 9 and 10. In regulation 2, several definitions are added, in particular for the purpose of the new chapter 4.

The introduction of an International Energy Efficiency (IEE) Certificate provides the need for amending chapter 2 of Annex VI (see also paragraph 14.1). In regulation 5, amendments require ships, to which chapter 4 of Annex VI applies to be subject to survey and certification taking into account the guidelines developed by the Organization. For that purposes guidelines on the survey and certification of the EEDI have been developed by resolution MEPC.214(63).

In the new chapter 4, regulations 19 to 23 are introduced into Annex VI. Regulation 19 identifies the application of chapter 4. Chapter 4 applies to all ships of 400 gross tonnage and above. Regulation 20 provides requirements on the attained energy efficiency design index (attained EEDI), where regulation 21 deals with the required EEDI.

For the method of calculation of the attained EEDI for new ships and for the calculation of reference lines for use with the EEDI, guidelines have been developed which were adopted on 2 March 2012 by resolutions MEPC.212(63) and MEPC.215(63), respectively.

Regulations 20 and 21 apply to new ships and to those having undergone a major conversion defined in regulations 2.23 and 2.24 respectively. Attained EEDI (regulation 20) shall be calculated for applicable ships as defined in regulations 2.25 to 2.35. The required EEDI (regulation 21) shall be determined for applicable ships as defined in regulations 2.25 to 2.31.

Regulation 22 provides requirements for the ship energy efficiency management plan (SEEMP). It is important to note that regulation 22 applies to new and existing ships. Each ship shall keep on board a ship specific SEEMP. This may form part of the ship's Safety Management System. The SEEMP shall be developed taking into account guidelines adopted by the Organization (resolution MEPC.213(63)).

In a new appendix VIII the form of the IEE Certificate has been added to Annex VI.

14.3 Actions by the marine administration

The marine administration is required to carry out the following actions in order to implement Annex VI:

.1 consider, in conjunction with other Administrations, where appropriate, requests for exemptions from certain aspects of Annex VI in order to facilitate trials of ship emission reduction and control technologies and, if the exemptions are granted, to monitor such trials with regard to any requests for extensions or with a view to withdrawing such exemptions (regulation 3.2);

.2 approve the on-site use of hydrocarbons that are produced in conjunction with activity on sea-bed minerals, which will then not be subject to the requirements of regulation 18 (regulation 3.3.2);

.3 approve and notify IMO of any equivalent fittings, material, appliance or apparatus or other procedures, alternative fuel oils or compliance methods used as an alternative to that required by the regulations (regulation 4);

.4 carry out initial, renewal and annual/intermediate surveys to ensure compliance with chapter 3 of Annex VI (regulation 5.1);

.5 carry out additional surveys as may be necessary (regulation 5.1);

.6 consider establishing appropriate alternative measures to demonstrate compliance for ships not subject to surveys under the Annex (regulation 5.2);

.7 delegate surveys, if necessary, and establish procedures to receive notifications of corrective action, and that if corrective action is not taken then ensure the certificate is withdrawn and that the appropriate authorities of the port State are notified immediately (see also chapter 23) (regulation 5.3.1 and 5.3.3);

.8 approve changes to equipment, systems, fittings, arrangements or material covered by survey (regulation 5.4)

.9 receive notifications from ships when an accident has occurred or a defect is discovered that substantially affects the efficiency or completeness of its equipment covered by MARPOL Annex VI (regulation 5.5)

.10 issue or endorse certificates following surveys, in the prescribed format (regulations 6, 7, 8 and 9);

.11 where authorised as the competent authority, undertake port State control inspections as necessary (regulation 10, regulations 18.7.2, 18.8.2 and 18.10);

.12 investigate possible violations of the requirements (regulation 11);

.13 implement controls on the emission of ozone-depleting substances (regulation 12.2);

.14 establish the acceptable forms of the Ozone-Depleting Substances Record Book (regulation 12.6);

.15 agree to the fitting of identical replacement diesel engines (regulation 13.1.1.2)

.16 agree to exemptions for certain marine diesel engines installed on ships engaged in domestic voyages (regulations 13.1.2.2 and 13.1.3);

.17 agree to the installation of a Tier II replacement marine diesel engine (regulation 13.2.2);

.18 agree to the installation of Tier II marine diesel engine in certain circumstances (regulation 13.5.2.1 and regulation 13.5.2.2);

.19 carry out, as set out in chapter 2 of the NO_x Technical Code 2008, or delegate, pre-certification surveys of marine diesel engines and, following such surveys, issue certificates in the prescribed format (see also chapter 23) (regulation 5.3.2 and regulation 13.8);

.20 establish the required forms of the fuel oil changeover record (regulation 14.6);

.21 notify IMO of any ports or terminals where the use of vapour emission control systems is mandated together with other required information and ensure that the shore-based elements are duly approved, operated safely and do not unduly delay ships (regulations 15.2 and 15.3);

.22 if required, approve vapour emission collection systems for tankers (regulation 15.5);

.23 approve Volatile Organic Compound Management Plans (regulation 15.6);

.24 approve shipboard incinerators permitted under regulation 16 (regulation 16.6.1);

.25 agree to exemptions for certain incinerators installed on ships engaged in domestic voyages (regulations 16.6.2);

.26 notify IMO of cases where reception facilities are not available together with other required information (regulation 17.2);

.27 notify IMO of cases encountered by ships under its flag where it has been found that reception facilities are not available or are alleged to be inadequate (regulation 17.3);

.28 promote the availability of compliant fuel oils in ports and terminals under its jurisdiction and notify IMO of same (regulation 18.1);

.29 receive notifications from ships unable to procure compliant fuel oil (regulation 18.2.4);

.30 consider cases where ships have on board non-compliant fuel oil due to non-availability of same and notify IMO of findings (regulation 18.2.1 – 18.2.5);

.31 if required, analyse the representative sample of fuel oil in accordance with the verification procedure set forth in appendix VI of MARPOL Annex VI (regulation 18.8.2);

.32 authorize appropriate authorities within its jurisdiction to establish and apply fuel oil supplier registration schemes (regulation 18.9);

.33 receive notifications from another Party where a ship is found to have received fuel oil that is non-compliant with the requirements of regulation 14 or 18 of MARPOL Annex VI (regulation 18.9.5);

.34 approve, in conjunction with other Administrations, where appropriate, alternatives to the bunker delivery note (regulation 18.11);

.35 waive the requirements to comply with regulations 20 and 21 (regulation 19.4);

.36 verify the attained Energy Efficiency Design Index (EEDI) (regulation 20.1);

.37 consider whether a new or existing ship that has undergone a major conversion shall be regarded as a newly constructed ship (regulation 21.2); and

.38 establish that each ship has on board a ship specific Ship Energy Efficiency Management Plan (SEEMP), that may form part of the ship's Safety Management System (SMS) (regulation 22).

14.4   Action by the ports

In ports and terminals where the use of vapour emission control systems is mandated that the shore-based elements of such systems are to be duly provided, operated safely and in a manner that does not unduly delay ships (regulation 15.3).

In respect of reception facilities, the port shall:

.1 assess requirements for reception facilities for ozone-depleting substances and equipment and materials containing such substances; and

.2 assess the requirements for reception facilities for residues from exhaust gas cleaning systems as they are brought into service and are installed on ships using the ports (regulation 17.1).

Regulation 17 states that each Party to Annex VI undertakes to ensure the provision of adequate reception facilities for the above materials from ships using its ports, terminals and, in the case of ozone-depleting substances and related equipment, ship-breaking facilities. This requirement is, however, subject to the availability of the industrial infrastructure necessary to handle such materials properly. In this respect reference is made to resolution MEPC.199(62), 2011 guidelines for the reception facilities under MARPOL Annex VI.

14.5   Guidance

Guidelines relating to aspects of Annex VI have been developed by IMO to assist in the implementation of Annex VI (see chapter 25 for a listing).

14.6   Summary of actions and requirements for implementing Annex VI

14.6.1 Shipowners or operators should ensure that:

.1 surveys as required by the Annex are arranged to be duly undertaken and facilitated;

.2 there is on board an IAPP Certificate as required or, for ships under 400 gross tonnage, the appropriate alternative;

.3 where exemptions or alternative compliance means are to be applied, they are duly approved or accepted by the Administration, that the relevant documentation is retained with same and that any relevant controls or restrictions are complied with;

.4 if applicable, procedures are in place to prevent the deliberate discharge of ODSs;

.5   if applicable, ODSs or equipment or materials containing same, when removed from the ship, are delivered to appropriate reception facilities and the receipts for such delivery are duly retained;

.6   if applicable, there is a duly completed ODSs Record Book of the required form;

.7   there are, on board, for each diesel engine subject to the NO_x controls, EIAPP Certificates to the appropriate Tier, except as may be allowed in the case of certain Tier III requirements. There must also be on board the approved Technical Files, together with any approved amendment documentation. That these marine diesel engines are installed and retained in a compliant condition and that, where the parameter check method survey procedure is used that there is a duly completed Record Book of Engine Parameters;

.8   Tier III marine diesel engines, where required to be installed, are duly operated as required within ECAs established for NO_x emission control;

.9   where a relevant approved method is commercially available, it is duly installed as required, and the marine diesel engine is thereafter maintained in a compliant condition and accompanied by the relevant Approved Method File, or, alternatively, the marine diesel engine is otherwise duly certified;

.10   the sulphur content of fuel oil as used is appropriate to the area of operation or other approved equivalent means applied;

.11   where the changeover of fuel oil is necessary in order to comply with the relevant ECA established for SO_x and particulate matter emission control requirements, it is duly completed prior to entry to, and commenced on exit from, the ECA in accordance with written procedures, and necessary data are duly recorded in the form as required;

.12   for applicable tankers, where the use of vapour emission control systems is required, they are fitted and operated as necessary;

.13   for tankers carrying crude oil, there is on board an approved and implemented VOC Management Plan;

.14   there is no incineration of prohibited materials, and that any incineration undertaken is only carried out in an appropriate incinerator or as otherwise allowed;

.15   duly certified incinerators are installed as required and that such units are operated, by trained personnel in accordance with the operating manual, which is retained on board;

.16   fuel oils are ordered to be in a compliant condition and best efforts are made, and duly documented, to obtain compliant fuel oil and that, where such fuel oils are not available relevant parties are advised, as required;

.17   fuel oil bunker delivery notes and representative samples are signed for, retained as required and made available for inspection to competent authorities as demanded, and where the bunker delivery note or the representative sample is not in accordance with the Annex requirements, it is duly documented and advice of same is made to relevant parties;

.18   the attained EEDI and the required EEDI are calculated for each new ship or after a major conversion as required under chapter 4 of Annex VI;

.19   for each ship a specific Ship Energy Efficiency Management Plan (SEEMP) shall be kept on board taking into account the new regulation 5.4 of Annex VI; and

.20   there is on board an International Energy Efficiency (IEE) Certificate for ships which fall under the requirements of chapter 4 of Annex VI.

For .18, .19 and .20 reference is made to resolution MEPC.203(62).

14.6.2 The marine administration is required to take the actions outlined in paragraph 14.3.

14.6.3 Ports should have adequate reception facilities taking into account the constraints outlined in paragraph 14.4.

15 Provision of reception facilities

15.1 MARPOL requirements for the provision of reception facilities

Five of the six Annexes to MARPOL have regulations requiring the provision of reception facilities. These are:

Annex I – Oil: in loading ports, ship repair yards, bunkering ports (regulation 38)

Annex II – Noxious Liquid Substances (NLS) in bulk: in ports and terminals an adequate reception facility needs to be present for cargo residues resulting from compliance with Annex II and in ship repair ports where repairs to NLS tankers can take place (regulation 18)

Annex IV – Sewage: ports and terminals in all areas and in special areas in particular when ports and terminals are used by passenger ships (regulations 12 and 12*bis*)

Annex V – Garbage: all ports handling ships in national and international trade (regulation 8)

Annex VI – Ozone-depleting substances together with equipment and materials (such as insulation foams) containing same: in ports, terminals, repair ports and ship-breaking facilities. Potentially, residues from exhaust gas cleaning systems as these are developed and enter into service: in ports, terminals, repair ports (regulation 17).

All of the regulations are similar in the way they begin, stating that: "The Government of each Party to the Convention (MARPOL) undertakes to ensure the provision of …". The wording then differs but basically states to undertake to ensure the provision of reception facilities adequate to meet the needs of ships using their ports or terminals.

It should be noted that the regulations state that the Government undertakes to *ensure* the provision of reception facilities.

This does not mean that the Government of a Party must provide the facility; it means, in practice, that the Government can require a port authority or terminal operator to provide the facilities. This is the most likely way to proceed and the specimen regulations in appendix 10 of this manual reflect this approach (see also paragraphs 6.5.1, 6.5.2, 6.5.4, 6.5.5.and 6.5.6).

It follows then, that ports and terminals should be aware of the needs of ships and arrange the provision of the necessary reception facilities before implementation of each Annex of MARPOL.

Whilst it is recognized that generally speaking, ports should identify their needs on a case-by-case basis, most ports need reception facilities for garbage (Annex V) and many for oily residues (bunkering ports, major traffic ports, oil terminals and refineries that load oil in bulk).

Reception facilities for garbage could comprise containers that could receive the ship generated waste, and transfer it to waste handling systems ashore.

Reception facilities for oil, however, will very often be somewhat more complicated due to the fact that most residue oil comes with large amounts of water mixed with it. It is, therefore, highly advisable to have some sort of treatment facility in order to separate oil and water. Figures 8 and 9 give some schematic and pictorial ideas of what that could entail.

System 1: Portable storage tanks

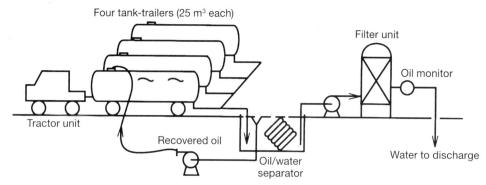

System 2: Fixed storage tanks

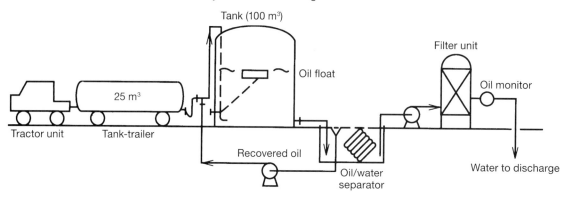

Figure 8 – *Systems for collection and separation of waste oil*

Ballast water reception system

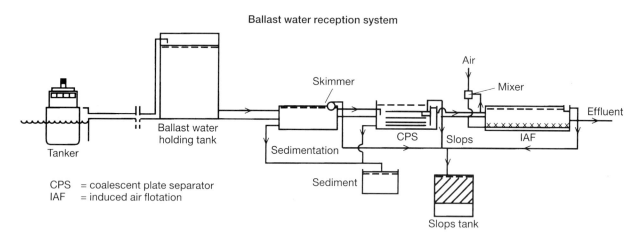

CPS = coalescent plate separator
IAF = induced air flotation

Flow path showing reducing pollution levels

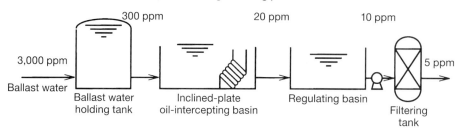

Figure 9 – *A system for the reception and treatment of ballast water*

Reception facilities for ozone-depleting substances (ODSs) and equipment or materials containing such substances (Annex VI) require adequate infrastructure to avoid the loss of these environmentally damaging gases to the atmosphere. In some instances gases can be banked and reused, however, in the case of the chlorofluorocarbons and halons, in view of their worldwide phase out since 2010 under the Montreal Protocol, this is not an option and instead requires controlled high temperature thermal destruction.

15.2 Comprehensive Manual on Port Reception Facilities

15.2.1 In order to promote the effective implementation of MARPOL, IMO has prepared the Comprehensive Manual on Port Reception Facilities which consists of fifteen chapters as follows:

1   Introduction

2   Legal background

3   Developing a waste strategy

4   National implementation

5   Planning reception facilities

6   Choice of location

7   Types and quantities of ship-generated wastes

8   Equipment alternatives to collect, store and treat ship generated wastes

9   Recycling ship generated wastes

10   Options for final disposal

11   Establishment and operation of reception facilities (including funding mechanisms)

12   Co-ordination and ship requirements

13   Options for enforcement and control

14   Small ships

15   Check list

Appendix A

15.2.2 The Comprehensive Manual on Port Reception Facilities (1999, second edition) provides a complete overview of the subject, including sections on law, planning and operations (chapters 1 to 5). In addition to dealing with the legal aspects and the planning of reception facilities, it also deals with the localization of reception facilities and types and amounts of wastes (chapters 6 to 8). The Manual covers the treatment of waste after it has been received; how to dispose of it (chapters 8, 9 and 10); and how to finance reception facilities (chapter 11). An important aspect is how to increase co-operation between the reception facilities and their users (the shipping industry) (chapter 12). An important aspect of the establishment of reception facilities is to ensure their use. Systems for inspection and control in respect of the adequate use of reception facilities is, therefore, increasingly important (chapter 13). The growing problem of reception facilities for small ships, local coastal ship traffic, and pleasure craft is also considered (chapter 14).

15.2.3 How to determine adequacy of reception facilities

MARPOL Annexes I, II, IV, V and VI call for adequate reception facilities. However, the interpretation of adequacy is left to the port State concerned and its users (the ships visiting its ports). Determination of adequacy has proven difficult. In order to provide guidance regarding the determination of adequacy, on 13 March 2000, MEPC has adopted the "Guidelines for Ensuring the Adequacy of Port Waste Reception Facilities" (resolution MEPC.83(44)). In this context, Governments were urged to meet their obligations to ensure the proper provision of adequate facilities and arrange for effective receipt of ships' wastes in their ports.

Governments were also urged to take the necessary steps to ensure that the planning and establishment of new facilities are achieved in accordance with these Guidelines.

All this information is of direct need to guarantee compliance with MARPOL, however, it appeared that the concern for inadequate reception facilities needed a further follow up. In this respect IMO developed The Guide to Good Practice for Port Reception Facility Providers and Users. This guide is intended to be a practical users' guide for ships' crews who seek to deliver MARPOL residues/wastes ashore and for port reception facility providers who seek to provide timely and efficient port reception services to ships. The guide was approved in July 2009 and issued under MEPC.1/Circ.671. In July 2013, the guide was reissued by means of MEPC.1/Circ.671/Rev.1. The guide is also available via the GISIS database via www.imo.org (see paragraph 3.11.7).

During additional debates it became clear that the obligations on providing adequate reception facilities raised concerns, in particular for Small Island Developing States (SIDS). In that respect, reference was made to resolution MEPC.83(44) in which it is identified that port waste management planning on a regional arrangement can provide a solution when it is undertaken in such a manner as to ensure that vessels do not have an incentive to discharge wastes into the sea.

To implement the principle of regional arrangements, however, amendments to Annexes I, II, IV, V and VI were necessary. These amendments were adopted on 2 March 2012 by resolutions MEPC.216(63) and MEPC.217(63), with an entry into force date of 1 August 2013. For the implementation of regional arrangements, guidelines were developed to provide guidance for the development of a Regional Reception Facilities Plan (RRFP), to assist Member States in specific geographic regions of the world in the appropriate and effective implementation of regulation 38 of Annex I, regulation 18 of Annex II, regulation 12 of Annex IV, regulation 8 of Annex V and regulation 17 of Annex VI of MARPOL (see MEPC.221(63)).

15.3 Action by the marine administration

The action by the marine administration depends on the division of responsibility in the Government concerned.

The flag Administration regulates the shipping industry, and its task is to ensure that the ships flying the flag of the State concerned are obliged to use reception facilities. According to MARPOL, the flag Administration should also establish reporting routines for reporting of alleged inadequacies to the appropriate port State and to IMO. In this respect reference is made to MEPC.1/Circ. 469/Rev.2 (27 June 2013) with the title: Revised Consolidated Format for Reporting Alleged Inadequacies of Port Reception Facilities.

The port State Administration (which might be under the same ministry or in the same entity, but might also be different) is obliged by the Convention to ensure that ports/terminals have established adequate reception facilities and that these are reported to IMO. It is recommended that they establish information systems to inform visiting ships about the reception facilities as well. With respect to reporting to IMO and information systems reference is made to GISIS (see paragraph 3.11.7). It is also highly advisable that they co-operate with the authorities responsible for waste handling and treatment to get a holistic national waste system and avoid a proliferation.

15.4 Summary of actions required

15.4.1 Port State authorities

- assess requirements, based on ships' needs;

- ensure that adequate facilities are provided;

- establish port State control routines and other ship control routines to ensure that local and international shipping are using the facilities; and

- inspect and license (if appropriate).

15.4.2 Marine administration

– monitor the provision of adequate facilities at ports and terminals for the reception of garbage;

– notify IMO of all cases where the facilities provided were alleged to be inadequate;

– follow up reports of alleged inadequacy;

– provide the mandatory reporting to IMO

– issue enforcement orders if necessary;

– prosecute in cases of non-compliance; and

– inspect and license (if appropriate).

15.4.3 Ports and terminals

– establish the necessary reception capacity through the establishment of reception facilities or through the use of contractors; and

– establish sustainable financing systems for reception facilities.

16 Implementing Protocol I: reports on incidents involving harmful substances

16.1 Brief explanation of Protocol I

Under Protocol I, the master or person in charge of any ship must report the particulars of any incident which entails the discharge or loss or probable discharge or loss into the sea of (see also paragraph 25.10):

.1   oil, as defined in Annex I of MARPOL, in excess of the permitted quantity and rate;

.2   noxious liquid substances, as defined in Annex II of MARPOL, in excess of the permitted quantity and rate; or

.3   harmful substances in packaged form, as defined in Annex III of MARPOL (marine pollutants defined in the IMDG Code).

The report should be made to the nearest coastal State and contain such details as:

.1   the identity of the ships involved;

.2   time, type and location of incident;

.3   quantity and type of harmful substance involved; and

.4   assistance and salvage measures.

Guidelines have been developed by IMO (Provisions Concerning the Reporting of Incidents Involving Harmful Substances under MARPOL (1999 edition)), and these should be followed by all concerned.

The purpose of this reporting requirement is to enable coastal States and other interested Parties to be informed of any incident giving rise to pollution or to a threat of pollution of the marine environment, so that appropriate action may be taken. Such action may include measures to prevent, mitigate or eliminate danger to their coastline or to related interests.

On 27 November 1997, the IMO Assembly adopted resolution A.851(20) on general principles for ship reporting systems and ship reporting requirements, including guidelines for reporting incidents involving dangerous goods, harmful substances and/or marine pollutants. Via this resolution all marine administrations are urged to ensure that ship reporting systems and reporting requirements comply as closely as possible with the general principles specified in the annex to resolution A.851(20).

As a consequence of the total revision of MARPOL Annex II by 1 January 2007, the MEPC adopted amendments to resolution A.851(20) to reflect amongst others the revised categorization system. MEPC, like MSC, received from the Assembly the function of adopting or amending performance standards and technical specifications referenced in MARPOL and other IMO instruments. Therefore Assembly resolution A.851(20) should be read together with resolution MEPC.138(53).

16.2 Actions and requirements for implementation of Protocol I

16.2.1 The marine administration should establish a ship reporting system and notify mariners of full details of the requirements to be met and procedures to be followed. This includes:

.1   nominating the shore establishment responsible for operation;

.2   issuing instructions for the relay of information to those nominated to receive it;

.3   instructing those nominated to receive reports on what action they should subsequently take;

.4   ensuring that the purpose of the reporting system is understood so that appropriate actions may be instigated;

.5   prosecuting those who fail to report such incidents; and

.6   notifying IMO of the arrangements made so these may be circulated to other Parties and Member States.

16.2.2 Shipowners or operators should issue appropriate instructions to the masters and officers of their ships on the necessity of reporting in the standard format and using the procedures laid down in the guidelines of resolution A.851(20), as amended via resolution MEPC.138(53). Should the ship be unable to report as required, the owner, charterer, manager, operator or agent is required to assume these obligations.

17 Implementing Protocol II: arbitration

17.1   Brief explanation of Protocol II

If the settlement of any dispute between Parties to MARPOL cannot be settled by negotiation, it may be submitted to an arbitration procedure as set out in Protocol II. This procedure starts with one of the Parties informing the Secretary-General of IMO of its application for an arbitration tribunal and the details of the dispute. The procedure is clearly laid down in the articles and requires no further explanation.

17.2   Settlement of disputes through marine administration and IMO

Disputes between Parties to MARPOL are rare. Differences of opinion on the application of MARPOL do occur, and these are usually dealt with by negotiation between the marine administrations of the Parties involved. Where necessary, the problem can be taken up by the Parties making submissions to IMO's Marine Environment Protection Committee for an agreed (unified) interpretation of a regulation, a resolution on a procedure or further work, or by proposing amendments to the regulations. The outcome is usually agreed by consensus.

18 Duties of shipowners

18.1   Most of the shipowners of States that are Parties to MARPOL will be conversant with and fully understand the requirements of the Convention and its Annexes. Many others, especially those whose flag State is not yet a Party to MARPOL and owners of ships in domestic trade may not be fully acquainted with the requirements.

18.2   When the Government of a State decides to ratify MARPOL, its shipowners should be advised; it is then in their interest to study the Convention and its implications for their ships. As described in earlier chapters of this manual, the requirements of the Convention, its Annexes and Protocols should be reproduced as national regulations, and it is with the technical and operational requirements a ship should comply. Certificates issued to ships trading internationally will, however, indicate compliance with the regulations of the appropriate Annex of MARPOL and should be accepted by other Parties (see paragraph 3.5).

18.3   Shipowners should ensure that:

> **.1**   their ships are designed, constructed (or modified) and equipped to the requirements;
>
> **.2**   their ships are surveyed by their marine administration (or by those with delegated authority – see chapter 23) and the surveys are kept up to date;
>
> **.3**   their ships are maintained in satisfactory condition and not modified without approval and with respect to the relevant MARPOL requirements; and
>
> **.4**   their ships' masters and crews are instructed and trained to comply with the relevant MARPOL procedures and that all operational requirements receive the correct treatment.

18.4   An indication of the shipowner's responsibilities under each Annex is given in several chapters in part IV of this manual.

19 Equipment requirements: the options

19.1 General

Ships of non-Parties to MARPOL trading internationally, and especially those trading with countries that have implemented MARPOL, may be expected to be designed, equipped and capable of operating to MARPOL requirements. This is usually indicated by the ship holding a "document of compliance" issued by the classification society which classes the ship. Such documents are usually accepted by the marine administrations of Parties to MARPOL in lieu of the certificates required under MARPOL, possibly with some port State inspection, where appropriate, in order to meet the "no more favorable treatment" requirement of article 5 (see paragraph 3.5). The owners of such ships are also likely to be generally aware of MARPOL requirements.

Ships which may be expected to require some modification and additional equipment in order to comply with MARPOL are those engaged in domestic trade and those engaged in international trade in a region where MARPOL has not yet entered into force. It may also be expected that the owners of such ships will not be fully aware of MARPOL requirements.

The design, construction and equipment requirements for ships under each Annex of MARPOL have been indicated in chapters 9 to 14 of this manual.

In the following paragraphs some comments in more general terms are made on equipment requirements and options for different types and sizes of ships with, where possible, the approximate cost of equipment to meet MARPOL requirements.

19.2 Annex I (Oil) – options for equipment requirements

Annex I affects all ships. All maritime States have ships, other than oil or chemical tankers, under their flag. Most countries import or export refined oils in tankers and many import and export crude oil. All ships, from the smallest to the largest, must comply with the discharge criteria of Annex I. The following paragraphs (19.2.1 to 19.2.6) cover some of the equipment requirements and options available for ships to comply with Annex I. There is no allowance given under MARPOL for ships reaching the end of their life or due for recycling. All must comply with the requirements, but sound judgment of the options available may limit the cost of compliance.

19.2.1 Compliance requirements for all ships

All ships above 400 gross tonnage need to be equipped with an approved bilge water separator or separator/filtering equipment capable of producing a maximum of 15 ppm oil-in-water effluent. Ships above 10,000 gross tonnage or ships which carry ballast water in empty bunker fuel tanks need to be equipped with high oil content alarm equipment to monitor 15 ppm effluent from separator/filtering equipment. This system shall also be equipped with an automatic stopping device so that any discharge that exceeds 15 ppm shall be stopped immediately. Ships below 400 gross tonnage may operate with retention of all oily water on board and discharge of all such oily water to a reception facility.

All ships need to be equipped with adequate holding capacity for sludge and oily residues generated on board. For ships below 400 gross tonnage this capacity must also include the bilge-water, unless the ship is equipped to the satisfaction of the Administration with equipment to purify the water and discharge it under Annex I criteria.

All ships need an international standard shore connection on deck, for pumping ashore bilge water and oily residues, and the related pumping and piping arrangements.

19.2.2 Compliance options for ships over 400 gross tonage

Ships of 400 gross tonnage and above need to be fitted with a 15 ppm standard bilge-water separator/filter. The separator/filter need not have a very large capacity. For most ships below a size of about 5,000 gross tonnage a capacity of 1 m³/h is adequate, minimizing the cost and problems of installation. There are many manufacturers and models of such separator/filters, approved to IMO performance and test specifications, on the market.

MARPOL includes provisions for a waiver for the bilge-water separator/oil filtering equipment, provided that the ship operates at all times within a special area and delivers all bilge-water ashore to reception facilities, the adequacy of which has been ascertained by the marine administration. Such limitations in the operation of the ships considerably restrict their trade possibilities and place a very heavy demand on reception facilities in many ports. All ships must have a standard shore connection on deck, connected by suitable pumping and piping arrangements to the waste oil tank and any bilge-water tanks.

19.2.3 Compliance options for ships below 400 gross tonnage

Ships below 400 gross tonnage should retain all bilge water on board and deliver this to reception facilities or, alternatively, be equipped to the satisfaction of the Administration with equipment for the purification of bilge water and discharging of the water under the Annex I criteria. As most ships of this size do not have any bilge-water collecting tanks, a practical method of achieving compliance would be to equip these small ships with some simple means, approved by the Administration, that ensures that the oil content of the effluent does not exceed 15 ppm whereby the oil can be kept on board and the purified water be discharged into the sea. The simple arrangements illustrated in figures 5 and 6 may be acceptable under national rules for these small ships and will satisfy this requirement. For very small ships, where even this simple system would be impracticable, other methods, such as manual removal of oil from machinery-space bilges with subsequent storage, may be used in order to ensure that the discharge criteria under Annex I are not exceeded.

19.2.4 Compliance requirements for oil tankers

Oil tankers of above 150 gross tonnage need, in principle, to comply with the following equipment requirements:

- an approved oil discharge monitoring and control system must be installed, able to monitor and record all discharges of oily water;

- the tanker must operate with a slop tank for the collection and settling of all oily water prior to discharge into the sea;

- the outlet of oily water must in principle be arranged above the ballast water line,

- however, exceptions are given in regulation 30 of Annex I;

- an oily-water interface detector must be available on board for the determination of the position of the interface between oil and water collected in a slop tank;

- additionally, the tankers must have equipment to handle engine-room bilge water as described in paragraph 19.2.2 for all ships.

A waiver may be given for the installation of the oil content monitor and the interface detector for tankers which operate at all times within a special area or is only engaged in voyages of less than 50 nautical miles from the nearest land outside a special area with a maximum duration of the voyage of 72 h and never discharge any oily water from cargo tanks into the sea. All contaminated ballast and tank washings must be discharged to reception facilities, and the Administration must be satisfied that these facilities are adequate for the operation.

Large tankers must also comply, as applicable, with requirements for crude oil washing, segregated ballast tanks or inert gas systems, etc. The necessary updating of the installations in a large tanker requires considerable experience if excess cost and complications are to be avoided and should only be considered by an operator with experience with such systems.

19.2.5 Compliance options for small coastal oil tankers

For small tankers which operate coastwise between ports and only occasionally sail to other ports in a limited area, the pattern of operations is ideally suited for utilization of the waiver possibility. These ships do not normally ballast any cargo tanks. (Should it be necessary, in extreme weather conditions, to take ballast into cargo tanks, this ballast must be pumped ashore). When the ships are due for dry-docking, tank cleaning will have to be done by collecting all tank washing water on board and discharging it, together with final cleaning of the tank used as the slop tank, at a reception facility. The oil companies controlling the trade should ensure that the ships which they engage comply with the regulatory requirements. The marine administration, when granting such waivers, must check that the operational arrangements of the ship are such that all oily water can and will be pumped ashore.

This arrangement will eliminate the cost of an oil discharge monitoring and control system installed in a tanker. This cost may be out of proportion for the small and normally relatively old tankers and can be eliminated by taking advantage of the waiver provision.

19.2.6 Equipment options

The ideal solution for handling engine-room bilge-water in all ships is thus the installation of 15 ppm bilge-water separator/oil filtering equipment and the necessary piping and pumping arrangements for pumping oily residues to the shore connection on deck. A suitable small tank in the engine-room must be arranged for collection of the oily residues. Every ship of 400 gross tonnage and above shall be provided with a tank of tanks of adequate capacity, having regard to the type of machinery and length of voyage, to receive the oil residues (sludge) which cannot be dealt with otherwise in accordance with the requirements of Annex I.

For ships below 400 gross tonnage, consideration may be given to the installation of a simple settling tank arrangement. A shore connection is required in parallel with the settling tank. This arrangement can be made on board, essentially without any equipment or shipyard cost, and will be considerably cheaper than the arrangement of a bilge-water holding tank of adequate size.

Consideration should be given to the option for tankers to operate with an equipment waiver for the oil discharge monitoring and control system and the other related equipment. Existing ships are assumed to have adequate piping to be able to perform tank washing with collection of all washwater in a slop tank. No additional equipment and installation costs are then required for small tankers operating in coastal trade, other than the cost of the engine-room installations mentioned above.

19.3 Annex II (Noxious liquid substances) – options for equipment requirements

Annex II affects only those ships which are certified to carry noxious liquid substances in bulk, that is in tanks that are fitted in, or are part of the ship's structure. An explanation of Annex II and its requirements has been given in chapter 10 and it will be recalled that ships certified to carry Category X and Y substances and in some cases Category Z substances will, be chemical tankers required to be certified to the IBC Code or the BCH Code (paragraph 10.6). Such ships should be equipped, tested and surveyed to meet Annex II (and SOLAS) requirements.

There are three options where exemptions on structural or pumping, piping and unloading requirements are allowed for ships certified to carry NLS:

    .1   ships whose constructional and operational features are such that ballasting of cargo tanks is not required under any condition and washing of cargo tanks is only required for repair or dry-docking, may be allowed to be exempted from the cargo tank stripping requirements subject to approval of design, construction, equipment, carriage of a restricted number of substances which are comparable and can be carried alternately in the same tanks without intermediate cleaning and previously arranged reception facilities when needed, all to the Administration's satisfaction.

.2 ships certified to carry individually identified vegetable oils (edible and animal oils included) as indicated by a footnote in chapter 17 of the IBC Code may be exempted from all the requirements for a ship type 2, including the maximum allowed quantity of 3,000 m^3 per tank, but only if the ship is identified as a ship type 3 and meets all requirements for a ship type 3, with the following cargo tank locations: double sides as per IBC Code and double bottom as per MARPOL Annex I. All operational requirements shall be subject to Annex II.

In this respect it should be carefully noted that, when any ship type 2 cargo, including a vegetable oil, is carried on board of a ship classified as ship type 2, all carriage requirements for a ship type 2 should be complied with. In any case, the carriage in excess of the maximum allowed quantity of 3,000 m^3 per tank is identified as a violation to the Convention.

.3 ships such as general dry cargo ships certified to carry individually identified vegetable oils as indicated by a footnote in chapter 17 of the IBC Code and off shore support vessels certified to carry identified NLS may be exempted from the carriage requirements only when the Administration has established measures based on guidelines developed by IMO. In these cases, references are made to resolution MEPC.148(54) for general dry cargo ships and to resolution A.673(16) regarding off shore support vessels which was amended via resolution MEPC 158(55).

Certain ships, especially coastal tankers and those on dedicated trades, may take advantage of these options in order to avoid some of the fitting, piping or stripping systems or other required equipment such as washing machines.

19.3.1 No indication of cost can be given within the scope of this manual for the fitting of underwater discharge outlet(s) for ships build before 1 January 2007 and certified to carry Category Z substances only, as this is very much a case for individual ship design and assessment. Most modern chemical tankers may be expected to be built to meet the requirements of regulation 12.6 and all chemical tankers engaged in international trade may be expected to comply with all requirements of the revised Annex II and the amended IBC Code.

19.4 Annex III (Harmful substances in packaged form) – equipment requirements

Annex III could be identified as a "transportation Annex" as it only affects the stowage, marking and labelling of products harmful to the marine environment carried in packaged form. This is the only Annex that has no requirements for reception facilities simply because there is no need to. An explanation of Annex III and its requirements has been given in chapter 11 of this manual. There are no equipment requirements under Annex III.

19.5 Annex IV (Sewage) – options for equipment requirements

Annex IV applies to all ships except those under 400 gross tonnage or certified to carry 15 persons or less (see regulation 2 of Annex IV). An explanation of Annex IV has been given in chapter 12 of this manual. For the options for equipment due note shall be given to MEPC.159(56), the Revised Guidelines on the implementation of effluent standards and performance tests for sewage treatment which apply to sewage plants installed on board on or after 1 January 2010.

It is necessary to consider ships' routes and ports before making a decision on the options, but many ships, especially larger cargo ships and most passenger ships, are already fitted with a sewage treatment plant. A number of these have been approved to IMO performance and test guidelines.

It should also be duly noted that during MEPC 64 in October 2012, the 2012 Guidelines on implementation of effluent standards and performance tests for sewage treatment plants were adopted by resolution MEPC.227(64). This resolution invites Governments to implement the 2012 Guidelines and apply them on or after 1 January 2016 and to provide the Organization with information on experience gained with the application of the 2012 Guidelines. MEPC 64 also agreed to a review of the nitrogen and phosphorus removal standard set forth in paragraph 4.2.1 of the 2012 Guidelines. This review should be undertaken by the Committee at its sixty-seventh session (second part of 2014) to determine that the required removal standards for nitrogen and phosphorus are met by type approved sewage treatment plants.

19.6   Annex V (Garbage) – equipment requirements

Annex V affects all ships, and an explanation has been given in chapter 13 of this manual. There are no equipment requirements under this Annex, but, for practical reasons, shipowners or operators need to make provisions for dealing with garbage on board ships. As indicated in paragraph 13.1, comprehensive guidelines have been published by IMO and these provide good advice on garbage handling, processing and equipment. Ship operators are advised to consider the following:

.1   minimization

.2   collection

.3   processing

.4   storage

.5   disposal – on board: incineration
                                        compaction
                                        comminution

.6   transfer to reception facilities

The methods and equipment chosen will depend on the type of ship and its service. Most of the processing equipment will be available as for use in shore establishments. With respect to incinerators, standards have been developed by IMO (see chapter 14).

19.7   Annex VI (Air pollution) – equipment requirements

Most equipment requirements of Annex VI apply to all ships irrespective of gross tonnage or size apart from certain aspects relating to the installation of Tier III diesel engines, see 14.2.2.

In view of the range of controls contained with Annex VI, see 14.2, each of the controlled pollutants is covered by a separate section below:

19.7.1   Ozone-depleting substances (ODSs) – regulation 12

The exclusion from the requirements of this regulation of sealed equipment without charging connections or removable components (this typically covers items such as small, domestic type, refrigerators, air conditioners and water coolers) should be noted.

In view of the pending worldwide phase out of chlorofluorocarbons (CFCs) and halons it is unlikely that new systems or materials using such gases have been installed on ships constructed on or after 19 May 2005 or that such systems or material are installed on existing ships on or after that date. However, where such installation has taken place, the following actions would be necessary:

In the case of refrigeration systems, where new or reused equipment using CFCs as the working medium have been installed then, in consultation with the suppliers of such gases, it should be possible to exchange the working medium to an appropriate alternative refrigerant.

Insulating foams blown with CFC would need to be removed and disposed of in an appropriate manner.

Firefighting systems containing halons. Manufacturers of firefighting systems should be consulted, however, it is probable that such systems, including distribution arrangements, will need to be refitted with systems working with an acceptable alternative medium. Disposal of the halon gas and associated equipment will need to be undertaken in an appropriate manner.

Portable firefighting equipment. In this context it should be noted that methyl bromide which historically was used in some portable fire extinguishers, is also an ODS and is included with the CFCs. Any such equipment will also need to be exchanged for other equipment using an acceptable alternative medium.

Systems, equipment and materials containing CFCs or halons not subject to prohibition under this regulation in view of the date clause may continue in service and be recharged or repaired as necessary, however, equipment and related procedures need to be in place to avoid the deliberate discharge of such gases to the atmosphere.

The pending requirements related to hydrochlorofluorocarbons should be considered when placing future contracts.

19.7.2 Nitrogen oxides (NO_x) – regulation 13

Marine diesel engines manufactured from the start of 2000 should have been produced in a compliant condition. For engines covered by the regulation requirements which do not have the necessary documentation (EIAPP Certificate and Technical File) but were produced from 2000 onwards, it should be possible to obtain retrospective certification; the engine builder's advice should be sought on this matter. In the case of such engines which have not been maintained in a compliant condition (NO_x critical components, settings or operating values) then these will need to be refitted or adjusted so as to comply. Where Technical Files have been lost or damaged over the years, the engine builder should be able to arrange the supply of approved replacement copies. EIAPP Certificates are to have been issued by, or on behalf of, the flag State of the ship onto which the engine has been installed; where this has not kept pace with transfer of flag then the engine certifier (see chapter 23) should be contacted.

Where diesel engines produced for other than the marine market (e.g., industrial) have been installed on board it may be found that the engine builder is not willing, or able, to provide the certification as required since the necessary emission test data and construction records do not exist. Under these circumstances retrospective certification is not just extremely difficult and potentially costly but may not be possible and, hence, it would be necessary to replace the engine; however, before so doing full discussions should be undertaken with engine builders, engine suppliers and engine certifiers in order to explore other possible options.

In the case of engines subject to "major conversion" on or after 1 January 2000, the certification status of such engines should be investigated. Where this is not as required the above scenarios should be considered as to the actions to be taken.

The pending requirements related to NO_x control (Tier II and Tier III) should be considered when placing future contracts for ships, engines or repairs.

19.7.3 Sulphur oxides (SO_x) and particulate matter – regulation 14

Where compliance is to be achieved by means of bunkering fuel oils not exceeding the appropriate sulphur limit value and a ship operates on different fuels when inside and outside an Emission Control Area (ECA) established for SO_x and particulate matter emission control, then it will be necessary to ensure that the storage, settling and service tank arrangements, together with associated transfer, treatment and service systems, are such that these different fuels can be maintained separately. Additionally, service systems to all fuel oil burning units must be capable of being changed over as required.

Fuel oil storage tank capacities and arrangements should be suitable for the intended operating areas taking into account ports where fuel oil is to be bunkered and the proportion of time to be spent within and outside an ECA.

In the case of the ECA requirements post 1 January 2015, when the sulphur limit value is decreased to 0.10% m/m, it must be anticipated that this will involve the use of distillate grade fuel oils. It should be ensured that all fuel oil burning units will be capable of operating with such grades of fuel.

In the case of compliance by other approved equivalent means the systems, equipment and controls appropriate to a particular arrangement will be required to be installed and maintained.

19.7.4 Volatile organic compounds (VOCs) – regulation 15

Where ports or terminals require certain tankers to be provided with vapour emission control systems these are to be in accordance with MSC/Circ.585, Standards for vapour emission control systems.

19.7.5 Shipboard incineration – regulation 16

Shipboard incineration is to be undertaken in an incinerator or as otherwise allowed under regulation 16.4. In the case of the incineration of polyvinyl chlorides this is only undertaken in incinerators certified to either MEPC.59(33) or MEPC.76(40).

In the case of incinerators installed on ships constructed on or after 1 January 2000 or units installed on existing ships on or after that date (other than as accepted for domestic shipping) the incinerator shall be type approved to MEPC.76(40), is to be provided with an operating manual which is to be retained on board and is to have reliable means to ensure that the temperature requirements given in regulation 16.9 are complied with as required.

Part V

Technical aspects of enforcement

20 Pollution detection and response

20.1 Overview

The purpose of this chapter is to provide guidelines to the investigating State on how to detect illegal discharges and secure evidence for the purpose of prosecuting violations of MARPOL. The Annexes I, II, IV, V and VI provide details on discharge requirements such as the minimum distance from the nearest land and the ship being *en route* maintaining course and a minimum speed. Reporting officials and other relevant individuals should pay close attention to these aspects of the discharge regulations. Violations can be broadly described as any acts which are in circumvention of pollution control and discharge provisions, except as expressly provided for (Exceptions) under the relevant regulations.

The key to effective enforcement of MARPOL is straightforward investigation work supported by laws that allow the basic collection and admissibility of legal evidence on all foreign and domestic ships within the State's jurisdiction. This evidence may include statements of witnesses, document (logs) review and analysis, photographs, and other related materials. Depending upon available resources, the investigator may have at his disposal evidence gathering techniques that run the gamut from simple to extremely complex. Since time is often the most serious limiting factor in pollution investigation cases, it is critical that a solid core of material evidence relating to the alleged violated discharge be established as quickly and accurately as possible. The simplest evidence gathering techniques (e.g., photography, interviewing witnesses, obtaining copies of documents, obtaining samples of the discharged material) are often the most expeditious ones, and should normally be employed before more sophisticated approaches are attempted.

Evidence gathering processes should include, where appropriate, standard procedures for: conducting searches; taking samples of material evidence; verifying the validity of certificates; interviewing witnesses; seizing and preserving physical evidence (including instructions for preserving the integrity of the chain of custody); training field investigators, and preparing investigative reports and forms.

20.1.1 Recording evidence: photographs

Evidentiary requirements are different in individual States and under different legal systems. Despite these differences, photographs are universally accepted as an excellent way of recording evidence because they document what was actually observed. Colour photographs are generally considered more persuasive than black and white photographs as they provide greater contrast between a pollutant discharge (e.g., an oil slick) and the background. Records of photographs should document the photographer's name, date, time, place (i.e., geographical co-ordinates), direction of shot, settings, and related information in order to assist enforcement efforts. Lighting may be a constraint in taking colour photographs. One should always be careful when using flash bulbs, electric lighting, or electronic equipment, as they can be a source of ignition around certain flammable atmospheres. If possible, when taking aerial photographs of a vessel trailing a suspected oil slick, care should be taken to ensure some photographs clearly show both the ship (preferably from an angle that "captures" the ship's name on the stern) and the slick in the same frame and a picture ahead of the ship with the ship's bow to prove that the ship is not sailing through an oil trace originating from a ship ahead.

20.1.2 Recording evidence: taking samples

Pollutant discharge sampling is another investigation technique that yields high quality and widely accepted material evidence. Often, a comparison of the physical and chemical attributes of a sample taken from a discharge with samples taken from the fuel and cargo tanks of nearby ships will pinpoint or, at a minimum, help identify the source of the discharge.

As pollutant sampling is a valuable evidence-gathering tool, serious consideration should be given to developing small portable sample kits for use by field personnel during investigations. Such kits should include, at a minimum, several solvent-washed sample jars with screw-capped teflon lids, small glass or pyrex funnels and adhesive labels to ensure security and protect the chain of custody. The addition of several fine mesh teflon screen segments and tweezers to the basic sampling kit will allow field investigators to collect "swish" samples of waterborne oil slicks that are too thin to collect with conventional grab sample techniques. Simple instructions covering the amount of oil to collect, sealing and security of samples, chain of custody procedures, and the delivery process to an analysis lab of samples should also be included.

The key to an effective sampling regime during any pollution investigation lies in a common sense assessment of the most likely sources of the discharge. Potential sources up-wind and up-current of the discharge should be given the highest priority. For oil discharges, the prioritization of ships for sampling should also be based on the type of oil discharged. For example, discharges of diesel oil should lead investigators to concentrate sampling efforts on nearby ships that use marine diesel oil as fuel or on tank ships that are carrying diesel oil as cargo. Conversely, discharges of bunker C oil should lead investigators to concentrate sampling efforts on ships that use heavy fuel oil or on tank ships that are carrying bunker C oil. Discharges that appear to be mixtures of used or "dirty" oils should lead the investigator to concentrate on the sampling of waste oils in the bilges of nearby ships. As waste oil mixtures in bilges are seldom homogeneous, investigators should concentrate their source sampling efforts on those locations closest to overboard discharge points in the bilge. In all cases, investigators assigned to discharge sampling duties should be well versed in the potential personnel hazards posed by pollutants and should take all reasonable precautions to guard against exposure to the substances involved.

20.2   Discharge, observation and investigation

Adequate knowledge of the behaviour of substances discharged into the water is critical to the successful detection and investigation of pollution incidents. Depending upon their specific gravity, chemical composition, and physical state, discharged substances will evaporate, float upon the water (positive buoyancy), mix and disperse into the water column (neutral buoyancy), or sink to the bottom (negative buoyancy). NLS may be miscible (dissolvable) in water – a characteristic that makes their illegal discharge difficult to detect. However, most oils discharged into the water will spread on the surface and float in an easily observable slick.

Oil slicks have several characteristics that simplify the investigator's task of determining the source of the discharge. For example, prevailing water currents and winds generally determine oil slick movement away from a discharge source. Under windless conditions, oil travels in the same direction and at the same speed as the surface current. Oil slicks are also affected by the wind and will travel downwind at approximately 3% of the wind speed. Thus, through vector analysis, field personnel can forecast the movement of oil slicks for response actions, as well as backtrack the path of oil slicks to determine the most likely source of the discharge.

While these general wind and current rules apply to oil slick movement on open waters, they are not always the controlling factors in slick movement on confined waters. For example, slick movement in harbours may be strongly influenced by tidal eddies, the wakes of transiting deep draft ships, piers and other manmade obstructions to current flow, and water discharges from shore-side facilities. These complicating factors make the backtracking of an oil slick's movement in a port extremely difficult. Moreover, investigators should realize that some ship berths act as natural collection points for oil slicks in many harbours. At these berths, investigators must avoid the tendency to automatically assume that the oil slick around the ship came from the ship in that berth. In these cases, an investigation of the ship's material condition (e.g., is there evidence of a recent oil spill on the ship's decks or oil staining on the hull?) and a review of the ship's recent operations (e.g., did the ship take on bunkers while at that berth?) may help determine if the discharge came from that ship. Investigators should sample the ship in the berth, as well as surrounding ships in the harbour, as potential sources of the discharge.

As mentioned above, tracing an oil slick's movement in an enclosed harbour or port is often a difficult and inexact process. In such cases, the investigator may find that the oil slick's characteristic collar and door provide a better clue to a discharge's source. For example, a discharge of diesel oil will usually form an amber or straw-coloured slick with a characteristic diesel odour. The logical source of such a discharge would be ships in the vicinity of the slick that carry diesel oil (as either cargo or fuel). Another common example involves discharges of gasoline, which forms a thin, rainbow coloured slick (or sheen) with an unmistakable odour. Investigators in this case should, of course, focus their evidence gathering efforts on nearby recreational ships that use gasoline as fuel or on oil tankers that carry it as cargo. Other types of oil exhibit unique colours, consistencies, and odours. Investigators should use these properties as clues in matching "fresh" discharges to their most probable sources. Investigators should note that oil characteristics will change with prolonged exposure to environmental factors (wind, sunshine, wave action).

Many noxious liquid substance, sewage, and garbage discharges lack distinguishing characteristics that could help link the alleged illegal discharge to the ship. Slicks of noxious liquid substances are usually colourless (though still visible), while garbage or sewage discharges from ship sources are often indistinguishable from garbage and sewage discharges from shore-side sources. For these pollutants, the proximity of the discharge to a certain ship is a good initial lead as to the potential source. However, a case of this type can only be successfully prosecuted if this lead is reinforced with solid material evidence (photographs, discharge samples, crew statements) that conclusively link the ship to the discharge. The revision of Annex V facilitates collecting evidence as in principle, all discharges are prohibited. There is for instance no longer a need to prove that the distance to the shore was violated.

20.3   Chemical analysis

Chemical analysis of oil and noxious liquid substance (NLS) discharge samples is a sophisticated technique for gathering/confirming evidence in a pollution investigation. Chemical analysis of pollutant samples can sometimes "fingerprint" the discharged oil or NLS and match it exactly with a source. However, in most cases, chemical analysis alone cannot conclusively prove that an oil discharge came from a particular ship. In most cases, the analysis results will only show whether or not the discharged oil is similar to the oil found on board one (or more) of the ships sampled. Sample analysis results of this "similar to the oil found on the ship" nature can be critical supporting evidence in cases where there is additional physical evidence (such as an oil stain on the hull of a ship outlining a path of discharge) that links the ship to the discharge.

Sample analysis relies on gas chromatography (GC) and mass spectroscopy (MS) techniques. Gas chromatography is less precise than mass spectroscopy in determining fine distinctions between similar chemicals. GC is also less expensive than MS and should therefore be used as an initial screening analysis to determine if any of the samples collected from various potential sources are "significant similar to" the discharged pollutant. If there is little or no similarity between the discharged oil and all the samples collected from the suspected source ships, then the sample evidence indicates that none of these ships is the source. Discharge samples demonstrating similarity to one or more of the suspected source ship samples during GC analysis should be further analysed with the more definitive MS technique.

As discussed above, sample analysis results of the GC and MS techniques usually state that the sample from the suspect source sample is "significant similar to" (with varying degrees of certainty) the discharge sample. Since sample analysis rarely provides a conclusive match between the illegal discharged pollutant and discharge source, it usually must be supported by other types of evidence gathered during the investigation. Moreover, chemical analysis technology is complex and its value in persuading a judge, jury or other adjudicative body that a violation occurred depends upon the ability of a scientist or other technician to explain the process used and its reliability. For this reason, it is recommended that enforcement agencies develop internal expertise or contract only with a limited number of businesses capable of providing the necessary analytical (and expert witness) services.

20.4   Remote sensing techniques

Remote sensing technologies provide another relatively sophisticated means of detecting and investigating pollutant discharges. Sensors can be used to detect floating oil or NLS slicks in cases where such slicks are not observable by the human eye due to darkness, cloudy weather, fog, and other visibility restrictions.

Most remote sensing devices used in pollution response operations are based either on radar or infrared sensor technologies. Radars provide oil and NLS slick imagery (radar returns) by measuring, comparing, and then displaying the differences in energy wave reflectivity exhibited by an oil or NLS slick and the surrounding waters. Infrared (IR) sensors provide images of oil and floating chemical slicks by differentiating between the temperature of the slick and the temperature of the surrounding waters. Additional remote sensing technologies include ultraviolet sensors and microwave radiometers. While these technologies can locate a potential oil or NLS/chemical slick, they cannot positively identify the type of substance floating on the water. In almost all cases, follow-up visual observations are needed to verify that the floating substance is indeed a pollutant (versus organic matter) and to determine what kind of pollution (i.e., type of oil or NLS) is involved. Thus, the true value of sensor systems lies in their ability to expedite both spill response and investigation operations by locating the pollutant discharge at night or in poor visibility conditions.

All of the sensor technologies described above can be integrated with aircraft navigation systems. Such airborne sensor systems are invaluable in displaying and "mapping" the position of potential oil/NLS slicks and suspect sources. However, it bears repeating that while all of these techniques may be used to detect floating substances, they are not capable of identifying the type or the quantity of the substances comprising the slick. As is the case with chemical analysis, remote sensing information gathered for a pollution investigation generally needs to be supported by other material evidence, including witness reports, photographs, and samples from the discharge and suspect sources.

20.5 Detection of air pollution incidents

Unlike with oil, noxious liquid substances, garbage or sewage related incidents, cases of non-compliance with the various requirements of Annex VI will not generally be detected by means of observing or assessing the particular discharge from the ship. The potential exception to this could be the remote sensing of the emitted exhaust gas plume for SO_x (SO_2) concentration – the higher the sulphur content of the fuel oil the higher the SO_x (SO_2) concentration in the exhaust gas stream – however, while technically possible, there would be the complicating factor of whether, in the overall plume, diesel engine and boiler exhaust gases (with different excess air ratios) had been mixed and it would still be necessary to establish which combustion devices, since it may not be all, had been using non-compliant fuel oil and the particular composition, in terms of sulphur content, of that fuel.

Instead, compliance verification (or in this case the detection of non-compliance) will, in almost all instances, be undertaken by examining the actions taken, or not taken, and procedures followed by the ship's crew together with the actual condition of relevant equipment or aspects such as fuel oil composition. It is the outcome of these investigations which will be the basis for assessing whether or not an air pollution incident has occurred. This inspection based approach is covered in detail in section 21.7 in respect of the various aspects covered by Annex VI.

Aspects such as the controls on NO_x emissions impact on the manner in which diesel engines have previously been operated and serviced while the SO_x and particulate matter controls affect established fuel purchase and/or chartering arrangements. These considerations highlight the need for enforcement to be backed by education and recognition that business-as-usual is not the way forward.

20.6 Reporting

Under article 8 and Protocol I of MARPOL, the master or other person in charge of a ship is required to report any incident involving a pollution discharge or threat of a discharge. MARPOL requires Parties to make arrangements to receive and process reports of these pollution incidents. States receiving reports are required to notify the Administration of the involved (or suspected) ship, as well as inform any other State which may be impacted by the pollution. This reporting requirement extends to other agencies of a Party that observe pollution incidents. In addition to the Parties required to report pollution by law, other members of the maritime community and the general public may use an established reporting system to report discharges of oil, noxious liquid substances or garbage.

An Administration's competent pollution authority should create and promulgate a standard reporting form to capture reported data on pollution incidents. The report form should be designed so that it includes questions that elicit basic pollution discharge information from a non-professional (a member of the general public), as well as questions that solicit detailed data of the type that might be provided by a professional responder. This form should provide space for a summary of the observation report and space for a listing of suspect sources of the discharge. The reporting network should include procedures to gather data from other ports or States.

The Administration should clearly identify the main point of contact(s) for receiving and responding to pollution reports. The Administration should proactively notify the maritime community and the general public of this point of contact(s) and of the responsibilities delegated to it. Ideally, the official contact point for pollution reports should also have the necessary equipment and facilities to initiate a response. Moreover, the point of contact for pollution reports should also be the source of historical records of incidents. These historical records can be used to assess the probability and relative risk of pollution incidents occurring at various coastal locations. These records can also be used to allocate and deploy pollution detection and investigation resources in a manner that most effectively deters illegal oil and noxious liquid substances discharges or air pollution. For easy reference, an itemized list of possible evidence of contravention of the MARPOL Annex I discharge provisions is given in appendix 18 of this manual.

In addition to those reporting sources noted above, reports of pollution may also be received from sources outside the Administration. For example, another ship, agency, or State may provide information of a ship discharging harmful substances. Such information should be corroborated with covert, remote observations while the ship is underway or at anchor or by targeting it thorough examination of its papers and equipment. In other instances, a confidential source may have reported the alleged illegal discharge, or may have information that leads to the identification of a ship responsible for one or more discharges. Such a confidant may be a crew member or passenger from the suspected ship. In such cases, steps must be taken to protect the confidentiality of the information source during and after the reporting and investigation phases. Since confidential sources often report discharges at the possible risk of losing their livelihoods, it is imperative that trained investigators commence response and investigation activities immediately. This immediate response helps maintain the confidentiality of the informant, while improving the quality of the evidence gathered for the violation case.

While many alleged illegal discharges are detected as a result of reports, pollution may also be discovered through routine or targeted patrols or on board inspections. Patrols are most effective when the area and timing of patrols is scheduled based on analysis of historical pollution sightings or incidents. This type of comparative risk analysis for patrol deployment or inspection purposes should also include provisions for high risk factors such as known ship bunkering activity, known location of "repeat offender" ships, and known location of ships preparing to depart port.

20.7 Prosecuting offences

In all cases leading to possible prosecution, investigators should follow a standard procedure and chain of custody policy to ensure the evidence is preserved correctly. All evidence obtained during the investigation, such as the remote sensing data, the witness report, the sample analysis, on board condition of relevant equipment, fuel oil sulphur content and other additional information, results in an official statement. This statement, according to article 8 of MARPOL, must be relayed to the Administration of the ship involved whenever violations are suspected, including when the alleged violation is observed outside the jurisdiction of the observing country.

States should also attempt to create and maintain an accurate and current database of all violations. It is helpful to categorize the violations based on:

.1 violator's identity;

.2 type of violations; and

.3 geographic location of the violation.

Such a database facilitates resource allocation decisions by identifying repeat offenders, frequent discharge locations, and the most common types of violations. This database could also be used to identify those ship operators for whom administrative sanctions, such as civil penalties, are not an effective deterrent. Sharing of this database information with neighbouring States will greatly facilitate and enhance regional enforcement efforts.

In prosecuting MARPOL violation cases, it may be necessary to present physical evidence, witness statements, and testimony from experts. In many cases, field investigators themselves will be called upon to present testimony. Therefore, it is important to develop field procedures that ensure the sound collection and preservation of physical and testimonial evidence. Full documentation of observations, witness responses, and field investigator comments is crucial to the successful processing of an alleged violation. In some cases, Administrations will need to identify experts on the various detection techniques. These experts must be available to testify at hearings or trials leading to sanctions for violations of MARPOL.

21 Strategies for inspection

21.1 General

21.1.1 A major part of port State control involves verifying that the ship's Convention certificates are in order and accurately represent existing conditions. An inspection under port State control is a spot check on the quality assurance of the flag State and the owner. Port State control should be seen as a "safety net". Routine inspections under port State control might vary widely depending on the type, age and maintenance standard of the ship and the experience of the surveyor or port State control officer (PSCO). Up to one hour may be needed just to check the ship's certificates. If there is a need for further inspections, based on clear grounds, additional time in relation to the severity of the deficiencies and discrepancies may be needed.

21.1.2 The qualification requirements for PSCOs for the conduct of inspections are detailed in Assembly resolution A.1052(27). Where fully qualified surveyors are not available, other suitably trained personnel may perform certain inspections on behalf of the port State party to the Convention. Junior members with specific training may conduct many aspects of the inspection under the guidance of a fully qualified PSCO. Under MARPOL, licensed engineers can check to ensure that equipment required under the various Annexes are properly in place and functional, such as the marine sanitation device, incinerators, the oily water separators, and the oil discharge monitoring equipment and that diesel engines are retained in a compliant condition. Checks should verify that these devices have no obvious unauthorized modifications. Licensed deck officers can conduct safety inspections of the deck and cargo spaces, paying particular attention to indications of hull damage, leaks from cargo or fuel areas, and proper markings in accordance with the reviewed and approved dangerous goods manifest. When an officer detects any discrepancies, they should be brought to the attention of the qualified surveyor or investigator for resolution.

21.1.3 As there must be clear grounds for believing that a violation has been committed for port State control purposes, investigation should have specific focus. A team approach is often useful, with a qualified PSCO in charge of junior or other specialized personnel. For instance, a chief engineer may be able to detect subtle anomalies in the Oil Record Book, the Ozone-Depleting Record Book, the Record Book of Engine Parameters or the fuel oil change-over records but a person specifically trained in techniques of questioning witnesses might also be valuable on the team.

21.1.4 Port State control on operational requirements is different from the work of the PSCO on hardware as indicated above which is based on article 5 of the Convention (see paragraph 3.5). When a PSCO visits a ship and has clear grounds for believing that the master or crew are not familiar with essential operational procedures the PSCO may inspect the ship on these aspects like asking the crew to test equipment described in the garbage management plan. Where an Annex to MARPOL identifies an operational requirement that is mandatory to be carried out, a PSCO is allowed to witness any such mandatory operational requirement under that Annex. For instance, the mandatory prewash under MARPOL Annex II. No clear ground is necessary in this respect. The rights and obligations for PSC on operational requirements are reflected in every Annex individually so there is no such requirement in the articles of the Convention like there is for PSC on hardware requirements (article 5).

21.2   Annex I

In setting priorities for a compliance strategy, the marine administration would need to have an idea of which ships have the highest potential for being in violation, or where a deficiency would be most significant. Some of the considerations for the marine administration are as follows:

21.2.1   Ships with machinery for propulsion or other task-specific activities

These have fuel and lubricating oils that may leak into the bilges of the machinery spaces. Leakage may originate for instance from oil changes, routine maintenance, and fittings loosening due to normal ship operations. A newer and/or well-maintained ship should have fewer leaks. A leak on "well-maintained" ships would more likely be cleaned up than allowed to stay in the bilge. Older ships and "poorly maintained" ships should be prime targets for inspections.

21.2.2   Oil tankers

Such ships may have to clean tanks during the ballast voyage if there is a change of cargo, or if ballast water had to be carried in cargo tanks. Between unloading one cargo and loading another on the ballast voyage, the ship will have to put the resulting tank washings somewhere. There may be the temptation to discharge these oily waters to sea, even in excess of allowances. Ships with SBT or on dedicated cargo runs are not likely candidates for illegal discharges. Therefore, older ships without SBT or CBT, and those which change cargo and need to clean tanks should be targeted.

An effective strategy of the standardized examination would be to inspect the bilges of the pump room, specifically during every ship inspection. The piping must be examined to verify that those ships required to have oil discharge monitoring equipment (ODME) do so, and that all ships equipped with ODME do not have piping which allows bypassing of this equipment. As piping is an expensive and "permanent" modification, it is more likely that the monitor might be electrically bypassed or disabled. This allows water from the ODME to be discharged overboard regardless of the oil content. An examination will usually reveal if someone has opened or tampered with the monitor, while legitimate maintenance will be recorded in the maintenance log. Inspectors should also carefully examine the Oil Record Book Part II and the ODME to check for any instances where ships attempt to allow bilge water from the pump room, with or without oil, to be pumped overboard.

21.2.3   Collecting evidence

During an inspection, PSCOs may take photographs and samples if there is a significant presence of oil in the bilges of the engine room on any ship and/or cargo pump room on oil tankers. This would alert the crew to the seriousness of the inspection, and the likelihood of being caught should oily bilge water be pumped overboard. To reduce costs, the pictures need not be printed nor the samples analysed unless there was a later alleged illegal discharge. It is relatively easy to tell if there is or was oil in the bilges from the oil staining. If the bilges are clean, but the rest of the engine room, machinery space or pump room is not well maintained, one may presume an illegal discharge. If there is excess oil in the bilges, the inspector should be empowered to require removal of the oil. This is especially necessary for small ships not equipped with oily water separators.

This attention to the condition of the bilges, especially after documentary evidence has already been collected, will act as a significant deterrent to an illegal discharge. If a discharge has occurred and a ship is a reasonable suspect, the bilges can be re-examined and the samples analysed if there is a discrepancy. If the ship has left port before the bilges can be re-inspected, the sample analyses may be sufficient to prove the case. The Bonn Agreement Guidelines for Sampling and Identification of Oil Spills should be considered when developing sampling plans or strategies. Having established a priority for choosing which ships to examine, the Administration may follow the general guidelines found at appendix 16 of this manual. If a ship is suspected of having discharged in contravention of Annex I, the investigators may follow the guidelines found at appendices 17 and 18 of this manual.

21.3   Annex II

Ships carrying Annex II cargoes in Category X or high viscosity or solidifying cargoes in Category Y are required to prewash those cargo tanks after unloading and prior to departure from the unloading port unless an exemption is given by the Government of the cargo receiving country. Essentially there should be less residue on board per tank and its associated piping than the quantity that is allowed to be discharged under Annex II.. For Category X substances, there shall be mandatory washings, including disposal to a reception facility in the unloading port. For Category Y, unloading shall be done in accordance with the P & A Manual to ensure that the residual quantity in the tank and its associated piping is not in excess of the regulations. Where it is not possible to achieve the residual quantity by efficient stripping (in case the product falls under the definition of solidifying and high viscosity substances) a prewash is required. The ship should discharge the washwater with the effluent into a reception facility at the port of unloading. Where a mandatory prewash is required and the Regional Reception Facility Plan is applicable to the port of unloading (see paragraph 15.2.3), the prewash and subsequent discharge to a reception facility shall be carried out as prescribed in regulation 13 of Annex II or at a Regional Ship Waste Reception Centre specified in the applicable Regional Reception Facility Plan.

At the request of the master of the ship, an exemption may be granted for the prewash subject to the following conditions:

.1   the tank will be reloaded with the same or a compatible cargo;

.2   the tank will not be washed or ballasted at sea and a written confirmation is given to the effect that the prewash shall take place at another port and the consequent washings shall be discharged to a suitable shore reception facility in that port; or

.3   cargo residues will be removed by ventilation in accordance with a procedure approved by the Administration.

The successful completion of the prewash of a product in Category X shall be witnessed and endorsed in the cargo record book (CRB) by a surveyor appointed or authorized by the Administration. For a prewash of a product in Category Y there is no requirement for mandatory witnessing of the operation by a surveyor. However, any operation that is witnessed, prewash or efficient stripping, as a consequence of a regular visit by a surveyor appointed or authorized by the Administration shall be endorsed in the CRB. Unlawful discharge, which may be the result of mechanical failure, human error or an intentional act, can be detected following an examination of the cargo inventory (paper trail afforded by the Cargo Record Book).

Ships certified to carry noxious liquid substances may find the need to wash the tanks in between the carriage of different cargoes for a commercial reason. The mandatory prewash shall take place in the unloading port and subsequent washings shall be discharged to a shore reception facility. Tanks that are unloaded from substances that do not require a mandatory prewash shall be unloaded in accordance with the P & A Manual. This operational procedure is called "efficient stripping". It is of major importance that the operational requirements, prewash or efficient stripping, carried out in port are monitored so that the ship can proceed to sea with an amount of residues per tank and associated piping which is not in excess of the quantities permitted under the Annex. Any water added to the tank after a prewash or efficient stripping with the aim to ballast the tank or commercially clean the tank may be discharged into the marine environment when the ship is *en route* with a speed of at least 7 knots, at a distance of not less than 12 nautical miles off the coast in a depth of water (charted depth) of at least 25 m. The discharge shall be made below the water line. These elements are important to control by the PSCO via the CRB, the ship's log and the nautical chart if the ship is suspect towards an alleged violation.

When inspecting ships carrying noxious liquid substances, such inspection might follow the general guidelines at appendix 16 of this manual. Surveyors should take all necessary safety precautions when entering potentially dangerous locations on ships. In the event that a ship is suspected of having discharged in contravention of Annex II of MARPOL, the investigators are advised to follow the guidelines found at appendix 17 of this manual.

21.4 Annex III

In the absence of any requirement for surveys and certification under this Annex, there are three elements for verifying the requirements of Annex III via the IMDG Code. These are:

.1   a detailed inspection of the required documentation;

.2   an on board verification that the paperwork matches the marking and stowage on board and that the labelling is correct; and

.3   whether the stowage is in accordance with the requirements of Annex III.

It is important to note that the vehicle for the implementation of MARPOL Annex III is the IMDG Code. This means that an inspection can never be only for Annex III (pollution hazard) or only for the IMDG Code (safety hazard). These two instruments go hand in hand and in the text below a logic mix is made.

A thorough review of the dangerous goods manifest can be a painstaking and protracted process. A novice could take hours verifying each entry with the Code without checking the general cargo manifest to ensure that none of the cargo listed there, should have been listed on the dangerous goods manifest. The task can be lengthy even for an experienced inspector who is familiar with the Code and with the transit pattern of routine cargo within the port. To conduct this inspection while the ship is in port may seriously delay the ship's departure. In general, it is a good practice to require advance copies of the dangerous goods manifest. This facilitates the agency's work and also minimizes delays to the ship. Customs or other officials who are involved, as a matter of course, in detailed reviews of goods entering and/or departing the country may be also required to check for the detailed requirements under the IMDG Code.

The other aspect of the enforcement is ensuring that the dangerous goods manifest correctly represents what is being carried on board. It generally does not take a long time to inspect the ship itself with dangerous goods manifest in hand to verify that dangerous goods on the manifest are stowed in the locations indicated, are properly marked, are packaged in good condition, and there are no marked dangerous goods are on board which are not listed on the manifest.

A critical element of safety for the ship and for the port is the proper segregation of dangerous goods. While stowage locations may be listed on the dangerous goods manifest, the average Customs official who is reviewing the documents may not sufficiently appreciate the requirements for segregation. Therefore, it might be useful for the on board inspection to verify that the stowage plan complies with the IMDG Code. Wherever a Customs official conducts the initial review of the advance documents, the reviewed documents should be submitted to the Administration for more informed review of the stowage plan.

As a specific element of enforcement of Annex III, the on-board inspector should be on the continuous look out for any indicators that there may have been an incident involving dangerous goods or harmful substances. During the on board inspection, key alarm indicators would include, but not be limited to, damp, crushed, or otherwise flawed packages; evidence of clean up, such as an area being unusually clean or damp; a large pile of rags or other material which might have been used to clean up a spill; members of the crew wearing unusual protective clothing; or evidence of burns or rashes. If the inspector has sufficient reason to suspect a spill, the master or person in charge may be asked specifically if there were any incidents. Crew members may also provide further details. The log should be also reviewed for any reports of incidents. In the event of a positive response, the inspector should ensure that all the required reports were made. Appropriate action as provided for in the Convention should be initiated if a violation is detected, whether or not it was reported.

21.5 Annex IV

All ships of 400 gross tonnage and above or certified to carry more than 15 persons are required to have some sort of sewage system on board. The Administration shall establish measures for ships which do not meet the above measures. Sewage systems tend to be expensive to maintain and might be subject to frequent failures. Reception facilities for sewage are available but in some places scarce and possibly expensive when used. As the discharge of sewage that is not disinfected or comminuted is permitted beyond 12 nautical miles from the nearest land, ships are likely to have piping in place to discharge directly overboard, with a valve to switch between the overboard discharge and the sewage system. Unless there is significant pressure and/ or monitoring capability by the port or coastal State to ensure that the valve is in the proper position and the system is kept in working order, pollution from sewage discharges is likely to occur. The only practicable strategy is to make inspection of the sewage system a key part of routine boarding. Such inspection would include reviewing the maintenance and repair logs for the system, verifying the amount of consumable goods if used, and ensuring the valve is secured in the proper position. The visual inspection of the system may give an indication of whether it is in good running order.

Notwithstanding the type of sewage system to be deemed appropriate for a particular ship, which may range from a holding tank to a treatment plant, inspections of the system should be conducted as a part of routine boarding. Unless there is intelligence to the contrary, ships would not be targeted solely for sewage inspections.

Such examination might follow the guidelines found at appendix 16. In the event of a discharge in contravention of Annex IV of MARPOL, the investigator might follow the guidelines found at appendix 17.

21.6   Annex V

All ships generate some kind of garbage and in this respect it should be noted that Annex V covers more than just garbage from the normal operation of the ship. It also includes loading and unloading excesses and spillages and cargo residues of solid bulk cargoes, where it should be recalled that such cargo residues are expected to be in small quantities that cannot be recovered using commonly available methods for unloading. The amount and types of garbage vary from one ship category to another. This makes tracking waste management practices difficult, however, the revision of Annex V greatly assists both compliance and enforcement. Regulation 10 of the revised Annex V addresses waste management plans and record keeping, which also facilitates enforcement by requiring documentation of waste handling practices and activities. Attention should be paid to inspecting records on board such as the Garbage Record Book.

Ships generate garbage in differing quantities as part of their normal operations. The highest end of the scale are cruise ships where at least three kilogrammes of garbage may be generated daily per passenger. In this way, the amount of garbage generated in one day by a cruise ship may be in the order of five or more tons. However, cruise ships tend to be equipped to handle high quantities of garbage and the crew is also trained for this purpose. The cruising yacht may be less able and less inclined to handle their garbage yield which has been averaged at two kilograms per person per day. With over 7 million ships in the world pleasure craft fleet, this is a significant source of garbage.

Waste management on ships is an important part of the operating budget, especially on cruise ships. Incinerators are costly to purchase, run and maintain, and have a history of frequent down time. Other factors such as the storage of wastes until discharging to a reception facility, or the discharge of food waste in the proper zone, and the need for garbage handling and separation are labour intensive. The crew might encounter a greater burden in complying with the provisions of the garbage management plans and the requirements of the Annex. While the majority of operators comply with Annex V, unfortunately, there may be the attempt by unscrupulous operators to cut corners and avoid costs by making illegal discharges. Enforcement officers should particularly target the Garbage Record Book to aid in detecting any unscrupulous practices.

The strategy for compliance with Annex V is mainly based on awareness of the crew. As discussed earlier in paragraph 5.10.3, also the involvement and awareness of the public in this area can be very useful in getting more violations reported, and in putting pressure to bear on consumer oriented operations such as the cruise lines.

In attempting to enforce the provisions of the revised Annex V of MARPOL, attention may also be paid to garbage that may have washed ashore. Sources of flotsam have been identified from markings, such as the ship's logo or name on the item. In addition, if large amounts of ship's garbage are observed and appear to be coming from a single source, a trajectory analysis may help in determining which ship(s) were in the area. However, the garbage found ashore is not only generated from ships. It is estimated that between 60% and 80% of the garbage concerned is land generated. Consequently, national public awareness and waste management programmes which target the shipping industry must also take account of the broader issue of the management of garbage from all sources. For this reason, enforcement of the revised Annex V of MARPOL, and indeed of all the Annexes, should be integrated with national legislation and enforcement programmes for garbage management and pollution control.

Unless there is direct photographic evidence, inspections and investigations into reports of discharges of Annex V material rely on detailed analysis of the waste stream of the ship, establishing what wastes should have been generated, what was properly "removed", and what is left. If there is a difference between the "removed" and "what is left", there might be a provable violation case. The level of interest shown in a detailed investigation can in itself constitute an education for the ship and a possible deterrent to illegal discharges. The guidelines at appendices 15 and 16 will be useful to the inspector when inspecting ships garbage management practices. The items listed below may also be adapted for use.

Suggested items for a check list for the Inspection of Ship Garbage Management Practices

1   Is there a garbage management plan in accordance with regulation 10?

2   Does it clearly identify the person responsible?

3   Does it address:

       minimization of wastes?

       recycling?

       separation of wastes?

4   Are there receipts for garbage landed ashore?

5   Is there a waste management training program?

6   What is the fate of plastics?

7   What equipment is used in waste management?

       compactor

       incinerator

       comminuter

8   Are there maintenance records of the equipment?

       does it work?

       is it properly maintained?

9   Is the plan followed, or if no plan is required, do the practices follow IMO Guidelines?

10   Calculate:

       the expected daily production of garbage (2 kg per person/day, on cruise ships
        3 kg per person/day);

       times the number of days at sea (since last landing of garbage ashore);

       subtract amount logged as incinerated, or otherwise disposed of in accordance with Annex V;

       estimate the amount of garbage on board;

       do the calculations balance? If not, why not?

11   Is there a suspicion of illegal discharge?

12   Can it be proved? If suspected but not provable, write a letter to the master reminding him of the regulations, of the penalties, and recommending use of shore-side reception facilities.

21.7   Annex VI

21.7.1   Each of the regulated pollutant streams controlled under Annex VI, see Chapter 14, have their own inspection procedure as outlined below with further details given in appendix 16:

21.7.2 Ozone-depleting substances (ODSs) – regulation 12

If no ODSs systems or equipment, as covered by regulation 12, are listed in section 2.1 of the Supplement to the IAPP Certificate then there should be normally no further need to consider this aspect. However, while it is most unlikely that systems or equipment containing chlorofluorocarbons or halons would have been retrospectively fitted, in view of the worldwide move to phase out such gases, it is possible that new hydrochlorofluorocarbon (HCFC) refrigeration systems or equipment have been fitted, as is acceptable until 1 January 2020, and if so should have been added to the Supplement listing.

Where systems or equipment containing ODSs are fitted a key point will be a well maintained, up-to-date, and readily available ODSs Record Book which accurately reflects the actions taken. As part of verifying the latter an inspector would look for indications, for example on marks on flanges or access covers of refrigeration compressors, that work has been done which has not been duly recorded.

Entries in the ODSs Record Book giving that gases or equipment have been landed to reception facilities should be supported by the relevant receipts. The key point here is whether such gases have not simply been vented to atmosphere and equipment disposed of in an inappropriate manner.

21.7.3 Nitrogen oxides (NO$_x$) emissions – regulation 13

All installed marine diesel engines potentially covered by the regulation 13 requirements should be listed in section 2.2 of the Supplement to the IAPP Certificate. Normally fitting an additional marine diesel engine is no small matter and therefore should be readily apparent, however, there can be the case where an additional engine is fitted outside the machinery spaces to meet increased power requirements, for example on a container ship where there is an increase in the number of refrigerated containers carried above the original design condition.

Detailed guidance as to how the inspection of the installed marine diesel engines in terms of NO$_x$ compliance may be undertaken is given in appendix 19 to this manual and is therefore not repeated here.

Generally, marine diesel engines installed on ships constructed before 1 January 2000 will not be subject to the NO$_x$ controls except where there is the requirement to apply an "Approved Method" or where an engine has been subject to major conversion, as defined, and it is on these points that an inspection should focus.

With regard to the major conversion clause, if a marine diesel engine is evidently substantially unchanged from its installed condition and the engine maintenance records show no upgrades to the fuel injection equipment, charge air system or principal combustion chamber components then it could be broadly accepted, within the terms of what can be achieved in the course of an inspection, that the engine has not been subject to a major conversion. If that is not the case the question, in the case of an uncertified marine diesel engine, is "why do these changes not represent a major conversion?", the concern here being that upgrades may have been applied without a full appreciation of the retrospective implications resulting from the Annex VI requirements.

21.7.4 Sulphur oxides (SO$_x$) and particulate matter emissions – regulation 14

This aspect essentially divides between compliance outside those ECAs established for SO$_x$ and particulate matter control and compliance inside such areas.

In the case of operation outside these ECAs, compliance will normally be demonstrated solely on the basis of the actual sulphur content of the fuel oils loaded as given on the retained bunker delivery note required by regulation 18.5.

Inside ECAs inspections should have a particular emphasis on ensuring compliance with the ECAs requirements. Consideration may need to be given to the guidance given by regulation 18.2, however, it should be expected that any relaxation of the ECA requirements would be applied only in exceptional cases and where thoroughly and robustly documented.

There will be those ships which never operate outside an ECA. For these the demonstration of compliance will again be solely on the basis of the respective bunker delivery notes. However, since ECA compliant fuel oil will generally be a higher cost product there will be a need to verify that duly compliant fuel has actually been delivered and hence used on board.

For those ships which operate both outside and inside these ECA the inspection issues can be divided into:

- – are the changeover procedures implemented as required and the necessary data duly recorded?
- – were ECA compliant fuel oils bunkered as indicated by the relevant bunker delivery notes?
- – were ECA compliant fuel oils actually used as required? This includes combustion devices which are remote from the main fuel oil treatment and service systems which have their own, free-standing, ready use tanks.

When an investigation is going beyond relying on the sulphur content, as given on the bunker delivery note, it requires analysis of the fuel oil in question; either the fuel oil as bunkered or in use depending on the point under investigation. Where the point in question is the fuel oil as supplied, the analysed sample would be the respective representative sample as required to be retained under the ship's control by regulation 18.8.1. In the case of the fuel oil in use, a specific sample would need to be obtained from the fuel oil service system, in which case the ship's engineers should be consulted as to an appropriate sampling point. The sample must be drawn from a position in direct connection with the oil actually in use and not from a position, such as a filter drain, where water or other materials will have accumulated. When sampling from a service system particular care is required as the oil will be under pressure and, in the case of residual fuel oil, at an elevated temperature; a typical sampling point would be prior to the booster pumps. In these instances, after having run an initial amount out into a waste oil container, a primary sample should be collected into a suitable, clean, metal container from which the sample for analysis is subsequently drawn That sample should be not less than 400 ml. and should be uniquely sealed with a label giving all relevant data; ship's name, IMO number, sampling location, date and time together with the inspector's details. The sulphur content analysis test method is ISO 8754: 2003, as given in appendix V of Annex VI. The procedure to be followed is given by appendix VI of Annex VI.

Where alternative approved equivalent methods are used to control SO_x and particulate matter emissions the means of undertaking inspections should be given within the approved accompanying documentation.

21.7.5 Volatile organic compounds (VOC) – regulation 15

In those cases where the use of duly certified vapour emission control systems is mandated the inspection requirement, in respect of air pollution control, would cover the verification that the necessary certification exists, that the system has been maintained in accordance with the approved arrangement and has been utilized as required.

For tankers carrying crude oil, an approved, and effectively implemented, VOC Management Plan shall be on board. Inspections will therefore necessarily focus on whether there is such a plan and, if so, that those responsible for its implication can demonstrate that it has been duly applied as required.

21.7.6 Shipboard incineration – regulation 16

Where shipboard incineration is undertaken it would need to be shown that no prohibited materials are incinerated. Questioning of those responsible for its operation would be a key point in verifying this point. The suitability of the incinerator for the disposal of polyvinyl chlorides would be indicated by the availability of one of the required Type Approval Certificate. If sewage sludge or sludge oil incineration is undertaken in main or auxiliary power plant or boilers it will be necessary to compare the operating record against the ship's position at those times to verify that the ship was outside ports, harbours or estuaries.

The necessary Type Approval Certificate should be available for those incinerators meeting the requirements of regulation 16.6.1. For these, the inspection should verify that the operating manual is available and that operators have been duly trained to implement the guidance as given therein, which should be both documented and verified by questioning. Additionally, the inspection should verify that the required temperatures have been achieved and duly monitored, and whether incinerators have been forced by attempting to load garbage before the unit was ready.

21.7.7 Fuel oil quality – regulation 18

In addition to those aspects which have already been considered related to regulation 14, the specific inspection points related to this regulation would be whether the required bunker delivery notes and representative samples have been retained as required.

In those cases where a bunker delivery note was provided but which did not give the required data as reflected in appendix V of Annex VI, or the representative sample was not drawn, handled, sealed or signed for on the part of the supplier in accordance with the relevant sections of the guidelines an inspection would verify whether such shortcomings have been duly documented and advised to the ship's flag Administration with copies to the bunker port authority and the fuel oil supplier.

21.7.8 Energy efficiency regulations – chapter 4

At the time of preparation of this manual guidelines for chapter 4 of MARPOL Annex VI are not available. The International Energy Efficiency (IEE) Certificate and the Ship Energy Efficiency Management Plan (SEEMP) could potentially be inspected.

Part VI

Organization

22 Basic marine administration

Before ratifying MARPOL, a State must be in a position to meet the requirements included in the articles and regulations. Almost all maritime States have accepted and implemented a number of existing international marine safety conventions (e.g., SOLAS, Load Lines, COLREG, etc.), and, in order to implement these, a marine administration in some form should exist. It is the existing marine administration that should first be examined when considering the undertakings and duties involved in ratifying MARPOL. The duties of a marine administration are shown in figure 10 and are applicable to a number of marine conventions. It is advisable to examine these duties in order to identify those required under MARPOL and decide how they will be covered. It will be seen that these duties have been divided into those of the "flag State", "port State", and "coastal State" for clarity, but there is inevitably a degree of overlap in these duties, especially with MARPOL.

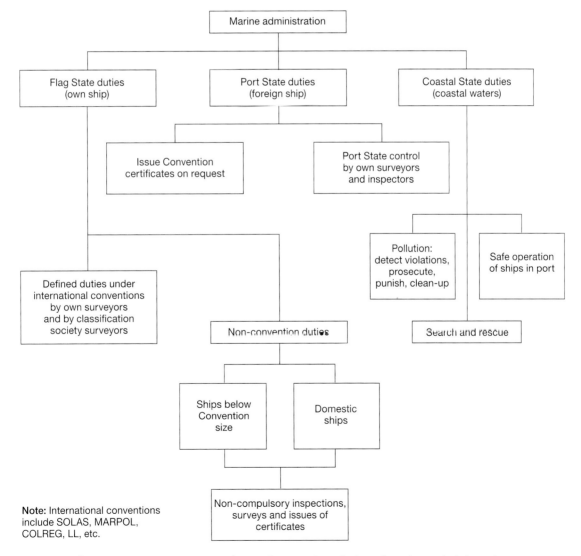

Figure 10 – *Convention and non-Convention duties of marine administrations*

22.1 A comprehensive marine administration

Many maritime States have large, well-organized and technically competent marine administrations. This is not intended to portray any particular marine administration and there will be differences of organization between States. The basic organization is, however, similar to many and may be readily adapted to illustrate any other existing comprehensive organization. The important features are:

.1 the link between a Government minister and a permanent administrative organization;

.2 the likely link with other transport systems;

.3 the link between shipping, ports and the organization that deals with marine emergencies (coast guard);

.4 the inescapable link between safety and the prevention and control of marine pollution;

.5 the professional and technical nature of the organization;

.6 the link with legal administrators; and

.7 the relationship between the headquarters (Administration) branch of the organization and the operational (surveys and inspections) branches.

22.2 Duties and implementation

The marine administration headquarters (Administration) and operational or district areas (survey and inspection) have distinctive roles in implementing MARPOL. The Administration and classification societies have in some cases a very close link.

22.2.1 Administrative duties – Headquarters

A short description of duties is given under each of the following:

.1 IMO representation: the successful functioning of IMO relies on the contributions made by Member States in the form of proposals, information, technical papers, reports, etc., and their participation in the meetings of technical committees. It should be understood that IMO itself does not set the standards or make the regulations but does provide the machinery to facilitate co-operation between Parties in producing agreed regulations, including dates on which they enter into force. It is, therefore, an important function of a marine administration to provide such contributions with respect to MARPOL in order to safeguard the marine environment, in the case of new developments, in a practical and economical manner whilst giving due consideration to the needs of all Parties and the problems involved. (Some Member States make significant efforts in this respect; others little or none.)

In all aspects, due consideration shall be given to IMO's Strategic Plan, the High Level Action Plan and the planned outputs in particular those related to the protection of the marine environment.

.2 Legislation: the work involved in preparing legislation to permit national regulations to be made to implement MARPOL is described in chapter 6.

.3 Regulations: these require preparation in a form suitable for national regulations (see chapter 6).

.4 Implementation of regulations: this requires the setting up of suitable organization (staffing at right level, accommodation, etc.).

.5 Instructions to surveyors: it is necessary, in most cases, to provide surveyors with national instructions and clarifying documents.

.6 Delegation of surveys and issue of certificates: no Administration has sufficient resources to deal with all the surveys and certification required by all the international marine conventions. This applies to MARPOL, and a certain extent of delegation is necessary. Decisions need to be made, including conditions, and the necessary organization has to be undertaken (see chapters 22 and 23 of this manual).

.7 Records of ship certification: necessary for control of flag State ships (ship particulars).

.8 Design approval: approval of ship design to meet regulations .

.9 Survey reports: confirmation that ships meet approved design and are constructed to relevant standards.

.10 Equipment approval: approval of required equipment in accordance with standards and guidelines.

.11 Issue of certificates: issue of certificates on completion of design approval and of surveys of ships and equipment.

.12 Violation reports: assessment of reports by own inspectors or other port State reports of contravention of the requirements of MARPOL.

.13 Prosecution of offenders: compiling of evidence and preparation of cases for prosecution.

.14 Monitoring reception facilities: necessary for the State to ensure adequacy of reception facilities.

.15 Informing IMO as required: provide details of arrangements for reporting incidents and on the items listed in article 11 of the Convention (see paragraph 3.11).

22.3 Surveys and inspections

A short description of the surveyors' and inspectors' duties is given in the following paragraphs:

.1 Ship survey to approved design: a most important element of compliance with MARPOL, which should be conducted by appropriately qualified surveyors (in association with headquarters).

.2 Inspections: inspections are necessary for port State control and to ensure compliance with required operational procedures.

.3 Investigations and prosecutions: surveyors and inspectors should be capable of investigating cases of non-compliance with constructional requirements, equipment requirements and violations. In association with headquarters (if necessary), they should arrange prosecution of offenders.

In this respect it is important to draw attention to the Harmonized System of Survey and Certification (HSSC). The development of this system received support from both MEPC and MSC. In March 1990, an amendment was adopted by MEPC to introduce the system in MARPOL. Meanwhile several Assembly resolutions were developed to introduce the HSSC system in mandatory IMO instruments. Through Assembly resolution A.997(25), adopted on 29 November 2007, the Assembly requested both MSC and MEPC to keep the survey guidelines under review and amend them as necessary.

As a follow up, MEPC developed in July 2009 by resolution 180(59), amendments to the survey guidelines under the harmonized system of survey and certification for the revised MARPOL Annex VI and agreed, at the 2009 Assembly, that these amendments were proposed for inclusion in the Survey Guidelines. As the Assembly had agreed that any amendment to the guidelines would be adopted at its next session, but that a consolidated version would only be issued after every odd session of the Assembly, the results of MEPC resolution MEPC.180(59) were incorporated in Assembly resolution A.1053(27), adopted on 30 November 2011.

22.4 Small marine administrations

Some States with relatively small fleets and small ports may not have the capability or competence to possess or acquire a comprehensive marine administration. They should, however, recognize their commitment to safety and pollution prevention and have a marine administration with the ability to fulfil their obligations under MARPOL (and other applicable conventions). In such cases, they are likely to depend heavily on a larger (associated) State and that State's marine administration, or on classification societies. An example of such a small marine administration is given in figure 20.

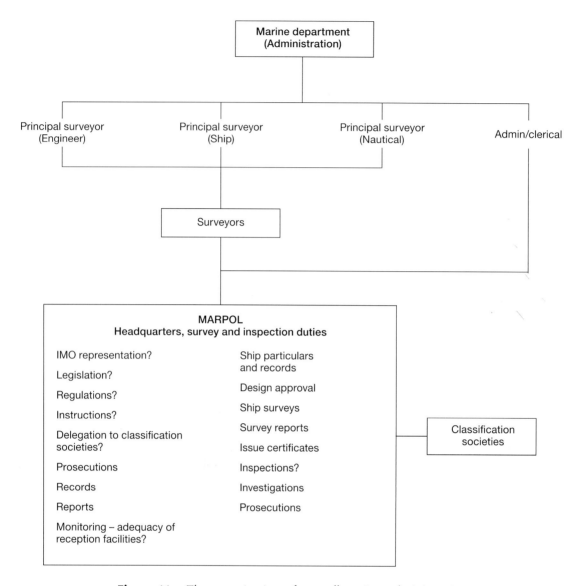

Figure 11 – *The organization of a small marine administration*

22.5 Composition of a marine administration

Factors affecting the composition of a marine administration are mentioned in paragraph 5.8 and chapter 24 and should be considered when planning for implementation of MARPOL. Duties as a "flag State", "port State" and "coastal State" have been outlined in figure 10 and the allocation of available resources between these duties may cause challenges. A large coastline with a large number of ports and, consequentially, a large number of ship visits may mean the provision of a large force of inspectors on port State control duties, even for a country with a small fleet. Other countries may have large fleets but small coastlines, with few ports requiring a large surveyor force for flag State duties. A balance should be achieved and, in general, surveyors may be employed as inspectors. As described further in chapters 23 and 24, delegation of certain duties is certain to be necessary.

22.6 Qualifications of staff

It is difficult to indicate a standard qualification for all of the staff of a marine administration. The fundamental requirement is that each grade should be capable of doing the job effectively at the level of the appointment. Given the international nature of shipping, this must involve comparison with similar appointments in both the individual's and other States. With these points in mind it may be useful to consider qualification requirements for professional administration, legal and survey staff. For example, on language skills, several regulations in the Annexes make references to the languages allowed to be used. For instance, the entries in the different record books might be made in the official language of the flag State; however, these entries shall also be made in English, French or Spanish. Therefore, proficiency in one of these languages to an appropriate level might be included within the assessment of qualifications.

.1 Administration staff: these are required at all levels in a marine administration. The staff should be qualified to a similar level as their counterparts in similar administrative organizations in their country and to a level appropriate to the other professional staff with which they will work in the marine administration.

.2 Legal staff: there will be comparatively few of these staff members, but they will, of necessity, be well qualified to meet the job requirements. This is likely to mean qualification in their own national law and in maritime law to master's degree standard, together with qualifications in international law and considerable working experience.

.3 Surveyors and inspectors: these should be qualified to degree standard in one of the three professional disciplines of marine engineering, naval architecture or nautical sciences. This should be coupled with service on ships at sea, or in shipyards, to gain several years' practical experience. Principal surveyors should have considerable experience in the field of survey or inspection and well-proven ability. In this area it should be recognized that the STCW Convention, which entered into force in 1984, is an attempt to establish minimum global standards for seafarers (which all countries are obliged to meet or exceed). Many surveyors and inspectors are likely to be drawn from such qualified seafarers and, as they will be inspecting and surveying the work of seafarers, should be qualified and have experience equal to or above the level of the most senior seafarers they will meet in the course of their duties. With regard to the entry into force date of the STCW Convention, it should be noted that major revisions to the International Convention on Standards of Training, Certification and Watchkeeping for Seafarers (the STCW Convention) and its associated Code have been adopted, known as the Manila amendments, with an entry-into-force date of 1 January 2012, including a five-year transitional period until 1 January 2017.

23 Delegation of duties by the marine administration

In the previous chapter it is stated that no Government has a marine administration extensive enough to perform all of its obligations under the international marine safety and maritime pollution prevention conventions. Delegation of duties is permissible under these conventions, but it should be recognized that this delegation does not relieve an Administration of its responsibilities.

23.1 Responsibilities of delegation

23.1.1 In respect of the MARPOL Convention, the responsibilities of delegation are clearly defined in the articles and regulations. Annex I (oil), Annex II (noxious liquid substances), Annex IV (sewage) and Annex VI (air pollution) all require survey and certification of the ship and each of these Annexes states the responsibilities of delegation in a similar form, as follows, under the headings of:

> **.1** Surveys
>
> "Surveys of ships, as regards the enforcement of the provisions of this Annex, shall be carried out by officers of the Administration. The Administration may, however, entrust the surveys either to surveyors nominated for the purpose or to organizations recognized by it.";
>
> "In every case, the Administration concerned shall fully guarantee the completeness and efficiency of the survey and inspection and shall undertake to ensure the necessary arrangements to satisfy this obligation.";
>
> and, under Annexes I, II, IV and VI:
>
> "An Administration nominating surveyors or recognizing organizations to conduct surveys as set forth in paragraph … of this regulation shall, as a minimum, empower any nominated surveyor or recognized organization to require repairs to a ship and carry out surveys if requested by the appropriate authorities of a port State."
>
> **.2** Issue or endorsement of certificate
>
> "Such certificate shall be issued either by the Administration or by any person or organization duly authorized by it. In every case the Administration assumes full responsibility for the certificate."

In the articles of the Convention (see paragraph 3.11.2), the Administration undertakes to provide IMO with a list of such nominated surveyors or recognized organizations, together with notification of the specific responsibilities and conditions of the authority delegated, for circulation to other Parties for the information of their officers.

It will be seen that the basic responsibility is for an Administration itself to carry out the survey and certification of its own-flag ships. It may, however, for pragmatic reasons, delegate these functions, subject to conditions. This requires a number of factors to be considered:

> **.1** what an Administration can or should do with its own resources;
>
> **.2** what is to be delegated;
>
> **.3** who, or which organizations, should be entrusted with delegated duties;

.4 what powers they should be given;

.5 how the completeness and efficiency of the delegated duties may be ensured;

.6 how arrangements may be made to provide IMO with the necessary relevant information; and

.7 how the assumption of full responsibility for certificates issued by others may be justified.

These factors are considered in the following paragraphs 23.2 to 23.11.

Note: In order to avoid confusion, in this chapter "surveys" refers to surveys for certification purposes under MARPOL and "inspections" refers to port State inspections of foreign ships and general inspections of the State Administration's own-flag ships.

23.2 What the Administration should do

A marine administration should, in general, have sufficient resources to carry out surveys of its own-flag ships and inspections of foreign ships in its ports by its own officers within its own State. However, where the marine administration does not have sufficient qualified surveyors or inspectors, or surveys of its own-flag ships have to be made outside the State, surveys, as well as the issuance of certificates, should be delegated, but only under the strict conditions permitted by MARPOL. Two other factors should be considered in deciding what should be done by the marine administration's own resources:

.1 The marine administration should have or acquire sufficient capability to carry out the responsibilities accepted under MARPOL. This means surveyors and inspectors, or other staff, who are qualified and experienced in survey and inspection duties and who are actively engaged in their work in order to keep up to date and maintain standards, to investigate incidents, and to carry out other duties mentioned in chapter 22.

.2 Port State inspections of foreign ships and general inspections of foreign-flag ships are normally performed by a marine administration's own inspectors. Classification societies are normally not used for inspection purposes.

23.3 What to delegate

After assessment, delegation as follows may be found advisable:

.1 surveys outside the State (where distance, cost, staff resources, etc., make this necessary);

.2 work which is well established and for which other organizations have the necessary expertise; and

.3 duties beyond the capability of the marine administration.

Under Annex II the Administration shall appoint or authorize surveyors for the purpose of implementing the measures of control under that Annex. The surveyor shall make an entry in the Cargo Record Book when he or she has verified that an operation has been carried out in accordance with the requirements as given in the Procedures and Arrangements Manual or has granted an exemption.

23.4 Which organizations or persons should be entrusted

The following are permitted under MARPOL but need to be assessed as outlined in paragraphs 23.6 and 23.8 to 23.11:

.1 "surveyors nominated" – this means individual persons who may or may not be members of an organization;

.2 "organizations recognized" – this, for all practical purposes, means the classification societies.

23.5 What powers should be given

Nominated surveyors and recognized organizations should have back up from the marine administration in order for them to carry out the duties required by MARPOL. In particular, power should be given to require repairs to a ship and to carry out surveys and inspections even though these may not be agreed or requested by the shipowner. The practical way is for individual surveyors to be formally given the powers of the relevant marine administration surveyor, or for the recognized organizations to be given the requisite power by national regulations or other formal authority, acceptable under national legislation.

23.6 Ensuring completeness and efficiency of duties

This can be done based on knowledge of the individual nominated surveyor or recognized organization. This may be backed up by requiring reports monitoring their work during inspections and by auditing the organizations.

23.7 Provision of information to IMO

It is a duty of the marine administration to provide a list of nominated surveyors and recognized organizations (classification societies) together with the specific responsibilities and conditions of the authority delegated (see paragraphs 3.11.2 and 23.1.1).

23.8 Assuming full responsibility for certificates

A marine administration should be satisfied that it can fully justify the assumption of this responsibility. Justification may be by similar means to those mentioned in paragraphs 23.6 and 23.9 to 23.11.

23.9 Appointment of nominated surveyors

In practice this may be done on the basis of a satisfactory assessment of each individual person based on his or her qualifications, experience and capability to conduct surveys to the required standard and the current MARPOL requirements. This is not an easy assessment to make nor for individuals to meet. It will require careful consideration by the Administration and some internal organization will be necessary to keep individual surveyors informed of changes to current requirements and to ensure that they are carrying out their duties completely and efficiently (see paragraph 23.6).

23.10 Recognized organization

In practice, the term recognized organization may be taken to mean a classification society and therefore, the term society is used in this paragraph. This does not preclude the use of other "organizations", and the contents of this paragraph would be relevant were any to be considered. An Administration should decide which societies it will entrust with the authority to act on its behalf for MARPOL purposes. These are likely to be the same as those authorized to act under other conventions, but they may be reduced or added to as necessary. Much will depend on the size of the flag fleet, the presence or otherwise of a national society which can meet its needs and the classification societies normally used by ships coming onto the State's register. It may not wish to limit its shipowners to one society, but the choice between a monopoly and competition, with possible reduction in standards of survey and enforcement, should be considered. It is essential that societies are clearly aware of the extent of delegation permitted. The Administration should give guidance in a written agreement which states that the societies are to survey to the full requirements of MARPOL. Clear instructions should be issued, laying down the action to be taken in the event of temporary non-compliance with the regulations, on the interpretation of regulations, on the issuing of exemptions where this is left to the discretion of the Administration, on the approval of equipment on behalf of the Administration, on the survey of ships not classed, and on the ready provision of information to the Administration when requested. With these points in mind, an Administration will wish to consider the service a society is prepared and able to provide.

23.11 Criteria that Classification Societies should meet

The general criteria to be met by societies acting on behalf of a marine administration should include the following:

.1   the society has to have the capability of setting and maintaining acknowledged technical standards;

.2   the society has to have sufficient experience and skill in performing technical surveys;

.3   the society needs to be represented worldwide, which requires a minimum number of personnel;

.4   the society should have published rules; and

.5   the society should be able to fulfil a continuing quality-assurance programme.

Such criteria are met by a number of the 50 or more classification societies which exist today. Ten leading Classification Societies are members of the International Association of Classification Societies (IACS), whose stated main purposes are:

.1   to promote improvement of standards of safety at sea and prevention of pollution of the marine environment;

.2   to consult and co-operate with relevant international and maritime organizations; and

.3   to maintain close co-operation with the world's maritime industries.

23.12 Guidelines

The International Association of Classification Societies (IACS) has been granted consultative status with IMO and in this capacity it has submitted a proposal containing principles to be included in new guidelines for delegating survey work to recognized organizations. Based on the proposal by IACS, the guidelines for the authorization of Organizations acting on behalf of the Administration were developed and adopted during the 18th session of the IMO Assembly via resolution A.739(18). This resolution has in its appendix the minimum standards for recognized organizations acting on behalf of the Administration.

Via this Assembly resolution, both the MSC and the MEPC were instructed to develop detailed specifications on the survey and certification functions of recognized organizations acting on behalf of the Administration. These specifications were adopted on 23 November 1995 via resolution A.789(19) where the Assembly urged Governments to apply the said specifications in conjunction with the annex to resolution A.739(18).

Information on this subject is given in appendix 12 of this manual.*

23.13 Summary of considerations and actions on delegation

The marine administration should:

.1   recognize responsibilities of delegation of surveys and certification;

.2   decide what duties should be delegated;

.3   decide on which persons to nominate, or which organizations are recognized and authorized to carry out MARPOL duties on its behalf;

.4   issue appointments or include authorization in national legislation; and

.5   inform IMO of action taken.

---

* Note that an updated and consolidated Code for Recognized Organizations (RO Code) was adopted by resolution MEPC.237(65) with the intention for this to take effect on 1 January 2015.

MARPOL HOW TO DO IT

24 Training of personnel

24.1 Consideration of training requirements for personnel

The need for training of personnel for the purpose of implementing the MARPOL Convention will depend on several factors and will need to be assessed by each State. This is a matter for the marine administration of a State and its shipping industry to explore. The following points should be considered:

.1 are the marine administration's own staff sufficiently conversant with MARPOL?

.2 are the staff of the marine administration technically competent to fulfil their obligations (see chapter 22)?

.3 do more appropriately-qualified staff need to be recruited and trained?

.4 are the shipowners conversant with MARPOL?

.5 what training do ships' masters and crew need?

24.2 Training possibilities

Having considered the above, the following possibilities for training might be considered:

.1 marine administration staff may visit other experienced marine administrations, as envisaged under article 17 of MARPOL (see paragraph 3.17);

.2 the technical competence of marine administration staff should be brought up to an adequate standard by training or recruitment or both;

.3 national seminars or courses (see paragraph 24.4) should be organized for surveyors, inspectors, administrators, lawyers, shipowners, masters and crew;

.4 regional training schemes might be arranged through IMO;

.5 attendance at the World Maritime University, especially for those capable of benefiting and subsequently returning to responsible positions in the marine administration and shipping industry;

.6 the marine administration may include MARPOL in the curriculum for seafarers' courses and examinations for certificates; and

.7 shipowners should ensure that shipmasters are aware of the required practical on board procedures and that they are carried out.

24.3 Pollution prevention and associated safety requirements

As indicated before, it is not possible to separate pollution prevention and safety. There are MARPOL requirements that are governed by SOLAS requirements – for example, crude oil washing systems, inert gas systems and chemical tanker operations. These all require specific training and certification of personnel.

24.4 National seminars and courses

The following points might be considered by those responsible for setting up local training for marine administration personnel and ships' officers:

24.4.1 Administration personnel

A training programme is necessary to make administrative and inspection personnel knowledgeable about the requirements of MARPOL and also to make surveyors competent in surveying ships for technical compliance with MARPOL. Inspection personnel must also be made knowledgeable about crude oil washing and the port inspection aspects of such activities. Further it is of the utmost importance that all involved stay informed on any amendments within one of the Annexes to MARPOL.

It is probably necessary that most of this information be conveyed in the national language. It seems necessary, however, to provide adequate information to local instructors in the first instance; some kind of combined training activity, in which experienced instructors initially work in parallel with local instructors, teaching courses for administrative and inspection personnel, would appear to be desirable. Such training should concentrate both on the content of MARPOL in total and on practical surveying procedures.

The timing for such training must be adjusted to suit the planned coming into force of the requirements in the ratifying State, so that sufficient time is given for thorough introduction to the practical requirements, but also that, the content of the instructions is not forgotten while the actual implementation is still being prepared.

When the schedule for the ratification and application of MARPOL requirements has been decided on, such training should be initiated, if necessary, with initial IMO assistance. It seems necessary to engage outside instructors to cover both theoretical and practical aspects of inspection. In view of the need for a mixed instruction programme, with alternating courses for administrative, inspection and ship's personnel, it may be necessary to provide such training assistance over a period of two to three months, possibly divided into several periods.

24.4.2 Ships' officers

Next to awareness on the protection of the marine environment, ships' officers need instructions about the requirements and regulations of MARPOL as a whole and instructions regarding the handling and operation of the categories of pollution-prevention equipment being installed on board ships in particular. For experienced officers this additional information could be given in very short courses of two to three days' duration. The training of instructors for this activity should be arranged in combination with the instructions mentioned in paragraph 18.3.4.

Should large tankers be brought under the flag of a ratifying State, it will be necessary to give the responsible officers the necessary training and for them to gain competence in the handling of crude oil washing, inert gas and related systems. This involves also the accumulation of the stipulated on board experience in operations of large tankers. The responsibility for the appropriate arrangements for such training must rest with the shipowner going into that business. The necessary competence can be obtained by a combination of using officers who have collected adequate experience abroad or foreign officers who have served for some time in the State's flag ships and, where necessary, participating in the special courses on crude oil washing and inert gas systems that are being arranged by some nautical training institutes in developed countries.

25 Guidelines, Codes and IMO publications relevant to MARPOL

A number of the regulations contained in the Annexes of MARPOL require procedures, equipment, construction, etc., to be based on guidelines developed by IMO or comply with IMO Codes. Some of these guidelines have been reproduced and exist as separate publications or are being produced. These are listed below. A complete listing of IMO publications is found on the IMO website (www.imo.org).

25.1 General

.1 *Particularly Sensitive Sea Areas* (2007 edition).

25.2 Annex I

.1 *Pollution Prevention Equipment* (2006 edition);

.2 *Dedicated Clean Ballast Tanks* (1982 edition);

.3 *Crude Oil Washing Systems* (2000 edition);

.4 *Inert Gas Systems* (1990 edition);

.5 *Condition Assessment Scheme* (2006 edition);

.6 *Enhanced Surveys for Bulk Carriers and Oil Tankers* (2008 edition);

.7 *Guidelines for the Development of Shipboard Marine Pollution Emergency Plans* (2010 edition). These Guidelines also cover the requirements in MARPOL Annex II.

25.3 Annex II

.1 *International Code for the Construction and Equipment of Ships Carrying Dangerous Chemicals in Bulk* (IBC Code) (2007 edition);

.2 *Code for the Construction and Equipment of Ships Carrying Dangerous Chemicals in Bulk* (BCH Code) (2009 edition);

.3 *Guidelines for the transport and handling of limited amounts of hazardous and noxious liquid substances in bulk on offshore support vessels* (2007 edition);

.4 *Revised guidelines for the provisional assessment of liquid substances transported in bulk* (MEPC.1/Circ.512, included in the IBC Code 2007 edition). The Guidelines provide step-by-step procedures for ascertaining the carriage requirements of all liquids for carriage in bulk.

25.4 Annex III

.1 *International Maritime Dangerous Goods Code* (IMDG Code) (2012 edition).

25.5 Annex IV

.1 *Revised guidelines on implementation of effluent standards and performance tests for sewage treatment plants (MEPC.159(55)).* [*]

25.6 Annex V

The revision of MARPOL Annex V, with an entry into force date of 1 January 2013, had major consequences on the *Guidelines for the Implementation of Annex V of MARPOL* which were published in the 2006 edition. Revised guidelines have been developed and adopted by resolution MEPC.219(63), 2012 Guidelines for the implementation of MARPOL Annex V. In addition the 2012 Guidelines for the Development of Garbage Management Plans were adopted by resolution MEPC.220(63). Both guidelines are included in *Guidelines for the implementation of MARPOL Annex V* (2012 edition). Refer also to the Amendments to the 2012 Guidelines for the implementation of MARPOL Annex V, as adopted by resolution MEPC.239(65).

25.7 Annex VI

.1 *NO$_x$ Technical Code 2008, resolution MEPC.177(58)*

This Code covers the survey and certification of marine diesel engines leading to the issue of an Engine International Air Pollution Prevention (EIAPP) Certificate. Also included are the means by which on-going compliance of a marine diesel engine is demonstrated over its service life, taking into account amendment and alterations which may be made to the marine diesel engine.

. 2 The following guidelines, as adopted by the reference resolutions, are applicable to Annex VI. Other guidelines may be developed on the basis of required need and therefore the latest IMO documentation should be consulted as necessary.

Guidelines adopted by resolution:

MEPC.180(59) Amendments to the survey guidelines under the harmonized system of survey and certification for the revised MARPOL Annex VI.

MEPC.181(59) 2009 Guidelines for port State control under the revised MARPOL Annex VI

MEPC.182(59) 2009 Guidelines for the sampling of fuel oil for determination of compliance with the revised MARPOL Annex VI

MEPC.183(59) 2009 Guidelines for monitoring the worldwide average sulphur content of residual fuel oils supplied for use on board ships

MEPC.184(59) 2009 Guidelines for exhaust gas cleaning systems

MEPC.185(59) Guidelines for the development of a VOC Management Plan

MEPC.198(62) 2011 Guidelines addressing additional aspects to the NO$_x$ Technical Code 2008 with regard to particular requirements related to marine diesel engines fitted with selective catalytic reduction (SCR) systems

MEPC.212(63) 2012 Guidelines on the method of calculation of the attained energy efficiency design index (EEDI) for new ships

MEPC.213(63) 2012 Guidelines for the development of a ship energy efficiency management plan (SEEMP)

MEPC.214(63) 2012 Guidelines on survey and certification of the energy efficiency design index (EEDI)

MEPC.224(64) Amendments to the 2012 Guidelines on the method of calculation of the attained energy efficiency design index (EEDI) for new ships

MEPC.231(65) 2013 Guidelines for calculation of reference lines for use with the energy efficiency design index (EEDI)

---

[*] Refer to the 2012 Guidelines on implementation of effluent standards and performance tests for sewage treatment plants (MEPC.227(64)) which supersedes these guidelines and should be implemented on or after 1 January 2016.

MEPC.232(65)   2013 Interim Guidelines for determining minimum propulsion power to maintain the manoeuvrability of ships in adverse conditions

MEPC.233(65)   2013 Guidelines for calculation of reference lines for use with the energy efficiency design index (EEDI) for cruise passenger ships having non-conventional propulsion

MEPC.234(65)   Amendments to the 2012 Guidelines on survey and certification of the energy efficiency design index (EEDI)

Many of the above have been consolidated into *MARPOL Annex VI and NTC 2008 with Guidelines for implementation* (2013 edition).

25.8   Reception Facilities

.1   *Comprehensive Manual on Port Reception Facilities* (1999, second edition)

.2   *Guidelines for Ensuring the Adequacy of Port Waste Reception Facilities* (2000 edition)

25.9   Port State control

.1   *Procedures for Port State Control 2011* (2012 edition).

Port State control has been an increasingly important feature in the field of maritime safety and protection of the marine environment over the past years, and several IMO resolutions have been adopted on this subject. These resolutions have been amalgamated into resolution A.1052(27), which was adopted by the IMO Assembly in 2011 and which constitutes a large part of this manual.

The resolution is intended to provide basic guidance on the conduct of port State control procedures and to afford consistency in the conduct of such inspections, in the recognition of deficiencies of a ship, its equipment and its crew, and in the application of control procedures.

25.10   Protocol I

.1   *Provisions Concerning the Reporting of Incidents Involving Harmful Substances under MARPOL* (1999 edition).

First published in 1986, this revised edition contains:

.1   article 8 of MARPOL;

.2   resolution MEPC.21(22);

.3   amendments to Protocol I of MARPOL – Provisions concerning Reports on Incidents Involving Harmful Substances;

.4   resolution A.851(20) – General Principles for Ship Reporting Systems and Ship Reporting Requirements, including Guidelines for Reporting Incidents Involving Dangerous Goods, Harmful Substances and/or Marine Pollutants;

.5   a list of agencies or officials of Administrations responsible for receiving and processing such reports.

25.11   Miscellaneous

.1   *Pollution Prevention Equipment Required under MARPOL* (2006 edition).

One of the objectives of MARPOL is to ensure that discharges from ships do not present an unreasonable threat of harm to the marine environment. Various resolutions have been adopted by the Assembly and by the Marine Environment Protection Committee (MEPC) of IMO that define or recommend suitable equipment, performance standards and procedures to achieve this objective.

The purpose of this manual is to provide easy and up-to-date reference to all applicable IMO resolutions on shipboard pollution-prevention equipment required under MARPOL. This includes equipment for the separation of oil from water, the treatment of sewage and the incineration of garbage and other shipboard wastes.

Appendices

Appendix 1
Example of a document of accession
(To be deposited with the Secretary-General of IMO, London)

WHEREAS the International Convention for the Prevention of Pollution from Ships, 1973, was adopted at London on 2 November 1973 by the International Conference on Marine Pollution, 1973,

AND WHEREAS the Protocol of 1978 relating to the International Convention for the Prevention of Pollution from Ships, 1973, was adopted at London on 17 February 1978 by the International Conference on Tanker Safety and Pollution Prevention, 1978,

AND WHEREAS . . . . , being a State entitled to become a party to the said Convention as amended by the said Protocol by virtue of articles 13 and IV respectively,

NOW THEREFORE the Government of . . . . , having considered and approved the said instruments, hereby formally declares its accession to the Convention as amended by the Protocol.

IN WITNESS WHEREOF I, ... , [President/Prime Minister/Minister for Foreign Affairs] of . . . . , have signed this Instrument of Accession and affixed [my/the] official seal.

DONE at . . . . this . . . . day of . . . .

[Seal]

[Signature]
[President/Prime Minister/
Minister for Foreign Affairs]

Appendix 2
Example of enabling legislation

. . . . may by Order . . . . make such provision as considered appropriate for the purpose of giving effect to:

1   the International Convention for the Prevention of Pollution from Ships (including its protocols, annexes and appendices) which constitutes attachment 1 to the final act of the International Conference on Marine Pollution (signed) (done) in London on 2 November 1973; and

2   the Protocol relating to the said Convention, which constitutes attachment 2 to the final act of the International Conference on Tanker Safety and Pollution Prevention (signed) (done) in London on 17 February 1978; and

3   any international agreement not mentioned in paragraph 1 or 2 which related to the prevention, reduction or control of pollution of the sea or other waters by matter from ships.

Appendix 3
Example of an "Order"

WHEREAS by virtue of [the enabling legislation reference] . . . . may by Order . . . . make such provisions as considered appropriate for the purpose of giving effect to:

[as paragraphs 1, 2 and 3 of appendix 2]

AND WHEREAS the Marine Environment Protection Committee of the International Maritime Organization by resolution . . . . dated . . . . adopted, in accordance with article 16(2)(d) of the Convention, amendments relating to Annex . . . . of the Convention;

AND WHEREAS the said Annex . . . . to the Convention and the amendments thereto come into force internationally on . . . . ;

WHEREAS this Order has been approved by . . . . ;

NOW THEREFORE, in exercise of the powers conferred by [enabling legislation], it is hereby ordered as follows:

1   The . . . . may make regulations for the purpose of giving effect to the said Annex . . . . to the Convention as amended.

2   Such regulations may in particular include provisions:

    (a)   with respect to approval of documents and the carrying out of surveys and inspections, and for the issue, duration and recognition of certificates and the payment of fees;

    (b)   that specified contraventions of the regulations shall be offences punishable . . . . by a fine . . . . on indictment by imprisonment for a term not exceeding . . . . . . years and a fine;

    (c)   for detaining any ship in respect of which such a contravention is suspected to have occurred.

Appendix 4
Example of regulations for the prevention of pollution by oil

The [. . . . .], in exercise of the powers conferred by [the enabling legislation or Order reference], hereby makes the following regulations:

Part 1: General provisions

1   Definitions

2   Applications and equivalents

3   Drawings, surveys and inspections

4   Certificates

Part 2: Control of operational pollution

5   Control of the discharge of oil

6   Segregated ballast tanks

7   Arrangement for crude oil washing

8   Existing product carriers of 40,000 tons deadweight and above

9   Requirements for tankers with tanks dedicated to clean ballast

10   Retention of oil residue on board

11   Oil discharge monitoring and control system and oily-water separating equipment

12   Tanks for oil residues, pumps, piping arrangement, etc.

13   Pumping, piping and discharge arrangements of oil tankers

14   Requirements for drilling rigs and other platforms

Part 3: Requirements for minimizing oil pollution from oil tankers due to side and bottom damages

15   Limitation of size and arrangement of cargo tanks

16   Watertight subdivision and stability

17   Double hull and double bottom requirements for oil tankers

18   Pump room bottom protection

Part 4: Oil Record Book

19   Ships which are to keep an Oil Record Book

20   The Oil Record Book

21   Entries in the Oil Record Book – responsibility

Part 5: Prevention of pollution arising from an oil pollution incident

22   Shipboard oil pollution emergency plan (regulation 37)

Part 6: Inspection, detention, penalties

Appendix 5
Example of regulations for the control of pollution by noxious liquid substances in bulk

The [. . . . .], in exercise of the powers conferred by [the enabling legislation or Order reference], hereby makes the following regulations:

Part 1: General provisions

1   Definitions

2   Applications

3   Categorization of noxious liquid substances

4   Drawings, surveys and inspections

5   Certificates

Part 2: Control of operational pollution

6   Discharge of noxious liquid substances

7   Measures of control

8   Arrangements for compliance with the discharge standards of noxious liquid substances

Part 3: Provisions for minimizing accidental pollution

9   Construction and strength

10   Ship Type requirements from a pollution prevention approach

Part 4: Cargo Record Book

11   The Cargo Record Book

12   Keeping of the Cargo Record Book – responsibility

Part 5: Inspection, detention, penalties

Part 6: Shipboard marine pollution emergency plan for noxious liquid substances

Appendix 6
Example of regulations for the prevention of pollution by harmful substances carried in packaged form

The [. . . . .], in exercise of the powers conferred by [the enabling legislation or Order reference], hereby makes the following regulations:

1   Definitions

2   Application

3   Prohibition of carriage of harmful substances in packaged forms

4   Packaging, marking, documentation and requirements in respect of stowage, etc

5   Inspection, detention, penalties

Appendix 7
Example of regulations for the prevention of pollution by sewage from ships

The [. . . . .], in exercise of the powers conferred by [the enabling legislation or Order reference], hereby makes the following regulations:

1   Definitions

2   Application

3   Surveys

4   Issue of certificate

5   Issue of certificate by foreign government

6   Form of certificate

7   Duration of certificate

8   Discharge of sewage

9   Exceptions

10   Standard discharge connections

11   Inspection, detention, penalties

Appendix 8
Example of regulations for the prevention of pollution by garbage from ships

The [. . . . .], in exercise of the powers conferred by [the enabling legislation or Order reference], hereby makes the following regulations:

1   Definitions

2   Application

3   Disposal of garbage within special areas

4   Disposal of garbage outside special areas

5   Special requirements for the disposal of garbage

6   Exceptions

7.1   Placards, garbage management plans and garbage record keeping

     .1   The Garbage Record Book

     .2   Keeping of the Garbage Record Book – responsibility

7.2   Garbage management plans

8   Inspection, detention, penalties

Appendix 9
Example of regulations for the prevention of air pollution from ships

The [. . . . .], in exercise of the powers conferred by [the enabling legislation or Order reference], hereby makes the following regulations:

Part 1: General

1   Application

2   Definitions

3   Exceptions and exemptions

4   Equivalents

Part 2: Surveys

5   Surveys

6   Certificates

Part 3: Requirements for control of emissions from ships

7   Ozone-depleting substances

8   Nitrogen oxides

9   Sulphur oxides and particulate matter

10   Volatile organic compounds

11   Shipboard incineration

12   Fuel oil quality

Part 4: Inspection, detention, penalties

Appendix 10
Example of regulations for the provision of reception facilities

The [. . . . .], in exercise of the powers conferred by [the enabling legislation or Order reference], hereby makes the following regulations:

1   Commencement and definitions

1.1   These regulations shall come into force on . . . .

1.2   For the purpose of these regulations, ... (definitions)

2   Application

2.1   These regulations apply to every port (defined to include port, estuary, haven, dock) authority or terminal operator of a port or terminal in [State].

3   Requirement to provide adequate reception facilities

3.1   The powers exercisable by a port authority or terminal operator in [State] shall include the power to provide reception facilities for:

> **.1**   residues or mixtures which contain oil or noxious liquid substances
>
> **.2**   sewage
>
> **.3**   garbage
>
> **.4**   ozon-depleting substances and equipment containing such substances when removed from ships
>
> **.5**   exhaust gas cleaning residues from exhaust gas cleaning systems from ships using the port or terminal.

3.2   Any power of a port authority or terminal operator to provide such reception facilities shall include powers to join with any other person in providing them . . . . (waste disposal authorities, contractors, oil industry, etc.).

3.3   A port authority in respect of its port or a terminal operator in respect of its terminal shall ensure that:

> **.1**   if the port or terminal has reception facilities, those facilities are adequate, or
>
> **.2**   if the port or terminal has no such facilities, adequate facilities are provided.

3.4   A port authority or terminal operator shall provide the [e.g., Department of Transport . . . .] with such information as it requires in respect of any reception facilities provided at its port or terminal.

4   Direction to provide adequate facilities

4.1   Where it appears to the [e.g., Department of Transport . . . .]

> **.1**   that the port or terminal has no reception facilities for (oil, noxious liquid substances, sewage, garbage, ozone-depleting substances or equipment containing such substances when removed from ships or exhaust gas cleaning residues) from ships; or
>
> **.2**   if the port or terminal has such reception facilities, that those facilities are not adequate: the [e.g., Department of Transport . . . .] may direct the port authority or terminal operator

to provide, or arrange for the provision of, such reception facilities as may be specified in the direction.

5 Use of reception facilities

5.1 A port authority or terminal operator or a person providing reception facilities may make a reasonable charge for the use of those facilities and may impose reasonable conditions in respect of the use thereof.

5.2 Any such reception facilities shall be open to all ships which, in the opinion of the port or terminal operator, are using the port or terminal for a primary purpose other than utilizing the reception facilities.

6 Penalties

Any port authority or terminal operator which fails to comply with any direction given under regulation 4 within the period specified in the direction shall be guilty of an offence punishable. . . . . by a fine . . . .

Appendix 11
Example of regulations for the reporting of pollution incidents

The [. . . . .], in exercise of the powers conferred by [the enabling legislation or Order reference], hereby makes the following regulations;

Commencement and definitions

1.1   These regulations shall come into force on . . . .

1.2   For the purpose of these regulations:

- *Air pollution* means any release of substances from ships into the atmosphere or sea which is within the definition of air pollution of the revised Annex VI of MARPOL.

- *Discharge* means any release, howsoever caused, from a ship and includes any escape, disposal, spilling, leaking, pumping, emitting or emptying, but does not include:

    .1   dumping, within the meaning of the Convention on the Prevention of Marine Pollution by Dumping of Wastes and Other Matter signed in London on 13 November 1972; or

    .2   any release for the purposes of legitimate scientific research into abatement or control of pollution.

- *The IMDG Code* means the latest consolidated edition of the International Maritime Dangerous Goods Code published by the International Maritime Organization, as amended from time to time by [. . . . .].

- *In packaged form* means in an individual package or receptacle, including a freight container or a portable tank or tank container or tank vehicle or ship borne barge or other cargo unit containing harmful substances for shipment.

- *Marine pollutant* means a substance which is identified as a marine pollutant in the IMDG Code.

- *Noxious liquid substance* means any liquid substance which is within the definition of noxious liquid substance in regulation 1 of the [Annex II – Regulations for the Control of Pollution by Noxious Liquid Substances in Bulk, 2007].

- *Oil* means petroleum in any form, including crude oil, fuel oil, sludge and oil refuse and any refined petroleum products, other than petrochemicals which are noxious liquid substances.

- *Sea* includes any estuary or arm of the sea.

- *Ship* means a vessel of any type whatsoever operating in the marine environment and includes hydrofoil boats, air-cushion vehicles, submersibles, floating craft and fixed or floating platform.

- *(Flag) Ship* means a ship which:

    .1   is registered in [State]; or

    .2   is not registered under the law of any country but . . . .

Application

2   These regulations apply to:

    .1   (flag) ships; or

    .2   other ships while they are within [the State] or the territorial waters thereof.

Duty to report

3.1   The master of a ship involved in an incident at sea involving:

   .1   an actual or probable discharge of oil or of a noxious liquid substance carried in bulk resulting or likely to result from damage to the ship or its equipment, or made or likely to be made for the purpose of securing the safety of a ship or saving life at sea; or

   .2   an actual or probable discharge of a marine pollutant in packaged form from the ship; or

   .3   an actual discharge during the operation of the ship of oil or of a noxious liquid substance in excess of the quantity or instantaneous rate permitted under the relevant provisions of Annex I (Oil) or Annex II (Noxious Liquid Substances), regulation … shall report the particulars of such an incident without delay and to the fullest extent possible in accordance with the requirements of regulation 4.

3.2   In the event of a report from such a ship being incomplete or unobtainable, the owner shall, to the fullest extent practicable, make or complete the report required by paragraph 3.1.

Contents of reports

4   The report (or the initial report, if there is more than one) shall in every case include:

   .1   the identity of the ship or ships involved;

   .2   the time, type and location of the incident;

   .3   the quantity and type of substance involved;

   .4   the assistance or salvage measures requested or being undertaken.

Supplementary reports

5   Any person required under regulation 3.1 or 3.2 to make a report shall, if possible:

5.1   make such a supplementary report or reports as may be appropriate in the circumstances,

   .1   supplementing the information contained in the initial report as necessary; and

   .2   providing information concerning further developments;

and

5.2   comply as fully as possible with any request for additional information made by or on behalf of the government of a State whose interests may be affected by the incident.

Report procedures

6   Reports shall be made by the fastest telecommunications channels available, with the highest possible priority, to the nearest coastal State.

Penalties

7   **.1**   Any breach of regulation 3 or 5 shall be an offence punishable on summary conviction by a fine . . . .

Appendix 12

Guidelines for the authorization of Organizations acting on behalf of the Administrations and specifications on the survey and certification functions of Recognized Organizations acting on behalf of the Administrations

Noting that the Administrations are responsible for taking necessary measures to ensure that ships flying their States' flag comply with the provisions of amongst others the MARPOL Convention, there was a need to develop uniform procedures and a mechanism for the delegation of authority to Recognized Organizations. By doing so it was also agreed that further minimum standards for the surveys and certification needed to be developed,

Based on the recommendations from both the Marine Environment Protection Committee and the Maritime Safety Committee guidelines were developed and finalized during the eighteenth Assembly in November 1993. These guidelines were issued as resolution A.739(18).

The Assembly further requested both MEPC and MSC to develop, as a matter of urgency, detailed specifications on the precise survey and certification functions of Recognized Organizations.

These specifications on the survey and certification functions of Recognized Organizations acting on behalf of the Administrations were adopted during the nineteenth Assembly on 23 November 1995. These guidelines were issued as resolution A.789(19) and urges Governments to apply the specifications in conjunction with the annex to resolution A.739(18).

Appendix 13
Extract from the 2008 edition of the IMDG Code

Information on the amendments to the marine pollutants provisions, which entered into force through amendment 34-08 to the IMDG Code on 1 January 2010. It should be noted that MSC 90, in May 2012, adopted an additional batch of amendments to the IMDG Code via resolution MSC.328(90), with a voluntary entry into force date of 1 January 2013 and mandatory from 1 January 2014.

It should also be noted that MARPOL Annex III, which also entered into force on 1 January 2010 by resolution MEPC.156(55), has been amended by resolution MEPC.193(61), also with an identical entry into force date of 1 January 2014. This appendix only reflects text of the amendments which entered into force on 1 January 2010.

1   Definition

Marine pollutants mean substances which are subject to the provisions of Annex III of MARPOL, as amended. These are substances, solutions or mixtures that are harmful to the aquatic organisms and the aquatic ecosystem of which they are a part and can be either in the form of liquids or solids.

2   General provisions

2.1   Marine pollutants shall be transported under the provisions of the revised Annex III of MARPOL, which entered into force on 1 January 2010. These provisions are included in amendment 34-08 to the IMDG Code.

2.1.2   Marine pollutants shall be transported under the appropriate entry according to their properties if they fall within the criteria of any of the classes 1 to 8. If they do not fall within the criteria of any of these classes, they will have to be transported under the entry: ENVIRONMENTALLY HAZARDOUS SUBSTANCE, SOLID, N.O.S., UN 3077 or ENVIRONMENTALLY HAZARDOUS SUBSTANCE, LIQUID, N.O.S., UN 3082, as appropriate, unless there is a specific entry in class 9.

2.1.3   When a substance, material or article possesses properties that meet the criteria of a marine pollutant but is not identified as such in amendment 34-08 to the IMDG Code, such substance, material or article will have to be transported as a marine pollutant in accordance with the Code.

3   Classification

3.1   A substance, material or article will be classified as a marine pollutant if the symbol **P** is indicated in the Index of amendment 34-08 to the IMDG Code, or in the absence of such an indication, if it meets the criteria of paragraph 2.9.3 of the IMDG Code. In this last case, it will remain the responsibility of the consignor to self-classify each substance, material or article. Therefore, the symbol "•"will no longer be assigned to generic and N.O.S. entries.

3.2   In amendment 34-08 there will no longer be any distinction between marine pollutant and severe marine pollutant (identified with the symbol **PP** in the Index of amendment 33-06). Therefore, substances, materials or articles previously identified as severe marine pollutants in amendment 33-06 to the IMDG Code will be identified as marine pollutant by the symbol **P** in amendment 34-08.

3.3   If a consignor has evidence that the substances, materials or articles no longer meet the criteria given in 2.9.3 of amendment 34-08 to the IMDG Code, but are indicated by the symbol **P** in the Index of this amendment, these substances, materials or articles need not be transported in accordance with the provisions of the IMDG Code applicable to marine pollutants, with the approval of the competent authority.

4 Marking of packages including IBCs and limited quantities

4.1 The marine pollutant mark required in amendment 33-06 to the IMDG Code will be replaced in amendment 34-08 with the mark as shown below. For packagings, the dimensions shall be at least 100 mm × 100 mm, except in the case of packages of such dimensions that they can only bear smaller marks.

Marine pollutant mark

Symbol (fish and tree): black on white
or suitable contrasting background

4.2 As a consequence of the deletion of the term "severe marine pollutant", the 500 g/500 mL inner packaging limitation, as provided in paragraph 5.2.1.6.1 of amendment 33-06 to the IMDG Code, has been removed. Therefore, packages containing all marine pollutants will be durably marked with the marine pollutant mark shown in 4.1 above, with the exception of single packagings and combination packagings containing inner packagings with:

    .1   contents of 5 L or less for liquids; or

    .2   contents of 5 kg or less for solids.

4.3 In addition, the limited quantity values applicable to substances, materials or articles formerly designated as severe marine pollutants have been harmonized with the provisions of the UN Recommendations on the transport of dangerous goods and SP944 has been deleted from amendment 34-08 to the IMDG Code.

5 Marking of Cargo Transport Units

5.1 The marine pollutant mark required in amendment 33-06 to the IMDG Code has been replaced in amendment 34-08 with the mark as shown in 4.1 above and will have sides of at least 250 mm × 250 mm.

5.2 The other marking provisions remain as they were before 1 January 2010.

Appendix 14
Extract from paragraph 2.9.3 of the 2008 edition of the IMDG Code

As indicated in appendix 13 to this manual, the 36-12 amendments of the IMDG Code (MSC.328(90)) and the amendments to MARPOL Annex III (MEPC.193(61), both with an entry into force date of 1 January 2014 affect the text of these instruments. This appendix only reflects text of the amendments in resolution MSC.262(84) which entered into force on 1 January 2010.

2.9.3 Environmentally hazardous substances (aquatic environment)

2.9.3.1 General definitions

2.9.3.1.1 *Environmentally hazardous substances* include, *inter alia*, liquid or solid substances pollutant to the aquatic environment and solutions and mixtures of such substances (such as preparations and wastes).

For the purposes of this section,

Substance means chemical elements and their compounds in the natural state or obtained by any impurities deriving from the process used, but excluding any solvent which may be separated without affecting the stability of the substance or changing its composition.

2.9.3.1.2 The aquatic environment may be considered in terms of the aquatic organisms that live in the water, and the aquatic ecosystem of which they are part.[*] The basis, therefore, of the identification of hazard is the aquatic toxicity of the substance or mixture, although this may be modified by further information on the degradation and bioaccumulation behaviour.

2.9.3.1.3 While the following classification procedure is intended to apply to all substances and mixtures, it is recognized that in some cases, e.g., metals or poorly soluble inorganic compounds, special guidance will be necessary.[†]

For definitions and data requirements a reference is made to the IMDG Code paragraphs 2.9.3.1.4 and 2.9.3.2.

In paragraph 2.9.3.3.1 it is mentioned that all substances shall be classified as "environmentally hazardous substances (aquatic environment)", if they satisfy the criteria for Acute 1, Chronic 1 or Chronic 2, according to the following tables:

Acute toxicity

Category: Acute 1	
96 hr LC_{50} (for fish)	$\leq$ 1 mg/L and/or
48 hr EC_{50} (for crustacean)	$\leq$ 1 mg/L and/or
72 or 96 hr ErC_{50} (for algae or other aquatic plants)	$\leq$ 1 mg/L

---

[*] This does not address aquatic pollutants for which there may be a need to consider effects beyond the aquatic environment such as the impacts on human health, etc.

[†] This can be found in annex 10 of the Globally Harmonized System of Classification and Labelling of Chemicals (GHS).

Chronic toxicity

Category: Chronic 1	
96 hr LC_{50} (for fish)	$\leq$ 1 mg/L and/or
48 hr EC_{50} (for crustacean)	$\leq$ 1 mg/L and/or
72 or 96 hr ErC_{50} (for algae or other aquatic plants)	$\leq$ 1 mg/L
and the substance is not rapidly degradable and/or the log $K_{ow} \geq 4$ (unless the experimentally determined BCF < 500)	

Category: Chronic 2	
96 hr LC_{50} (for fish)	> 1 to $\leq$ 10 mg/L and/or
48 hr EC_{50} (for crustacean)	> 1 to $\leq$ 10 mg/L and/or
72 or 96 hr ErC_{50} (for algae or other aquatic plants)	> 1 to $\leq$ 10 mg/L
and the substance is not rapidly degradable and/or the log $K_{ow} \geq 4$ (unless the experimentally determined BCF < 500), unless the chronic toxicity NOECs are > 1 mg/L	

Appendix 15

Classification flowchart outlining the process to be followed for marine pollutants

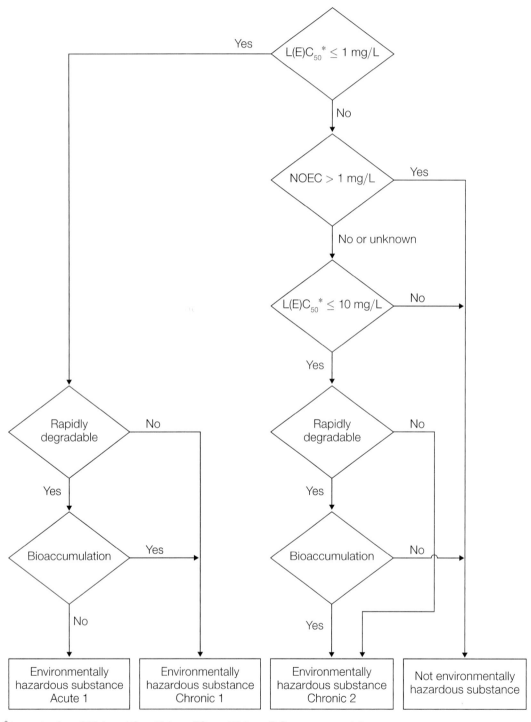

* Lowest value of 96-hour LC_{50}, 48-hour EC_{50} or 72-hour ErC_{50}, as appropriate.

Regarding the classification of mixtures, reference is made to paragraphs 2.9.3.4 and 2.9.3.5 of the IMDG Code (Amdt. 34-08).

Appendix 16
Inspection of certificates and relevant documents

1   Ships required to carry certificates

1.1   On boarding and introduction to the master or responsible officer, the inspector should explain the purpose of the inspection. In this explanation, the inspector should make it clear that the master or responsible officer remains accountable for the safety and operation of the ship and for any violations which may be caused during the inspection. With this in mind, the master should assign a knowledgeable officer to accompany the inspector to ensure that no tests are made which will risk the safety of the ship or risk a violation. To this end, the inspector should refrain from conducting tests himself, but rather observe the tests as performed by the ship's personnel or by the equipment specialist hired by the ship. The above notwithstanding, the inspector should be ever mindful of safety issues and of actions which might cause a violation. If there is any doubt as to the safety or wisdom of a test or action proposed, the inspector should delay said test or action until these concerns have been satisfied.

1.2   If the certificates are valid and the inspector's general impression and visual observations on board confirm a good standard of maintenance, the inspection should generally be confined to reported deficiencies, if any.

1.3   If, however, from general impressions or observations on board, the inspector has clear grounds for believing that the condition of the ship or its equipment does not correspond substantially with the particulars of one or more certificates, a more detailed inspection should be considered.

2   Ships of non-parties to the Convention and other ships not required to carry one or more certificates

2.1   As this category of ships is not provided with one or more certificates, the inspector will need to satisfy him- or herself with regard to the construction and equipment standards relevant to the ship on the basis of the requirements set out in the applicable Annex of MARPOL.

2.2   In all other respects, the inspector should be guided by the procedures for ships referred to in section 1 above.

2.3   If the ship has some form of certification other than the recognized certificate, like for instance a document of compliance, the inspector may take the form and content of this documentation into account in evaluating the ship.

3   Control

3.1   In exercising control functions, the inspector will be guided by established policies of the Administration, as well as his professional judgment, in determining whether to detain the ship until any noted deficiencies are corrected or to allow her to sail with certain deficiencies which do not pose an unreasonable threat of harm to the marine environment. In doing this, the inspector should be guided by the principle that the requirements contained in the Annexes of MARPOL in respect of construction and equipment and the operation of ships are essential for the protection of the marine environment, and that departure from these requirements could constitute an unreasonable threat of harm to the marine environment.

3.2   Specific guidelines for actions under port State control are found in resolution A.1052(27), "Procedures for Port State Control". These guidelines should be taken into consideration by the Administration in establishing national policy to deal with violations found in port. The established policy must be officially documented and available to the inspector.

A   The IOPP Certificate, its Supplement, the Oil Record Book and the shipboard oil pollution emergency plan

1   The inspector should examine the International Oil Pollution Prevention (IOPP) Certificate, including its supplement (Form A for the engine room of all ships and Form B for oil tankers), the Oil Record Book (Part I for the engine room of all ships and Part II for oil tankers) and the shipboard oil pollution emergency plan.

2   The IOPP Certificate carries information on the type of ship and the dates of surveys and inspections. As a preliminary check, it should be confirmed that the dates of surveys and inspections are still valid. Furthermore it should be established whether the ship carries an oil cargo and whether the carriage of such oil cargo is in conformity with the certificate.

3   If the inspector, from general impressions or observations on board has clear grounds for believing that the condition of the ship or its equipment does not correspond substantially with the particulars of the IOPP Certificate, a more detailed inspection should be conducted.

4   The inspection of the engine-room should begin with forming a general impression of the state of the engine-room, the presence of traces of oil in the engine-room bilges and the ship's routine for disposing of oil-contaminated water from the engine-room spaces.

5   The inspection of oil tankers should include the cargo tank and pump-room area of the ship and should begin with forming a general impression of the layout of the tanks, the cargoes carried, and the routines of cargo slops disposal.

6   Next, a closer examination of the ship's equipment as listed in the supplement to the IOPP Certificate (Form A for the engine room and Form B for tankers) may take place. This examination should also confirm that no unapproved modifications have been made to the ship and its equipment.

7   Every oil tanker of 150 gross tonnage and above, and every ship of 400 gross tonnage and above other than an oil tanker, must have an Oil Record Book part I (machinery space operations). Every oil tanker of 150 tons gross tonnage and above must also have an Oil Record Book part II (cargo/ballast operations). The inspector should make sure that the ship has an Oil Record Book on board in form as specified in appendix III of Annex I of MARPOL, as amended. The inspector should also make sure that the Oil Record Book is filled in according to instructions given in appendix III of Annex I of MARPOL as amended. In this respect also MEPC.1/Circ.736/Rev.2, issued on 6 October 2011 should be taken into account. This guidance is intended to facilitate compliance with MARPOL requirements on board ships by providing advice to crews on how to record the various operations in the ORB Part I by using the correct codes and item numbers in order to ensure a more uniform port State control procedure.

8   Every oil tanker of 150 tons gross tonnage and above, and every ship other than an oil tanker of 400 tons gross tonnage and above, shall carry on board a shipboard oil pollution emergency plan approved by the Administration. The Inspector should make sure that the ship is provided with such a plan.

9   Should any doubt arise as to the maintenance or the condition of the ship or its equipment, then further examination and testing may be conducted as considered necessary. In this respect reference is made to the IMO Survey Guidelines under the Harmonized System of Survey and Certification, resolution A.1053(27) adopted on 30 November 2011 (see also paragraph 22.3 of this manual).

B   Certificate of Fitness, Noxious Liquid Substance (NLS) Certificate, Cargo Record Book

1   The inspector should examine the Certificate of Fitness or the Noxious Liquid Substance (NLS) Certificate and the Cargo Record Book.

2   The certificate includes information on the type of ship, the dates of surveys and a list of the products which the ship is permitted to carry.

3   As a preliminary check, the certificate's validity should be confirmed by verifying that the certificate is properly completed and signed, and that required surveys have been performed. In reviewing the certificate, particular attention should be given to verifying that only those noxious liquid substances which are listed on the certificate are carried and that these substances are in tanks approved for their carriage.

4   The Cargo Record Book should be inspected to ensure that the records are up to date. The Book should show that the ship left the previous port(s) with residues of noxious liquid substances on board not in excess of those allowed to be discharged into the sea. It could also have relevant entries from the appropriate authorities in the previous ports. If the examination reveals that the ship was permitted to sail from its last unloading port under certain conditions, the surveyor should ascertain that such conditions have been or will be adhered to. If the surveyor discovers an operational violation in this respect, the flag State should be informed by means of a deficiency report.

5   If the inspector, from general impressions or observations on board, has clear grounds for believing that the condition of the ship, its equipment, or its cargo and slops handling operations do not correspond substantially with the particulars of the certificate, or are not carried out in accordance with the Manual a more detailed inspection should be conducted.

6   The more detailed inspection should begin with a further examination of the ship's approved Procedures and Arrangements (P & A) Manual. This Manual shall be in the format as specified in appendix IV to MARPOL Annex II.

7   It should include the cargo and, where applicable, the pump-room area of the ship and form a general impression of the layout of the tanks, the cargoes carried, pumping and stripping conditions and cargo.

8   Next a closer examination of the ship's equipment as shown in the P & A Manual may take place. This examination should also confirm that no unapproved modifications have been made to the ship and its equipment.

9   Should any doubt arise as to the maintenance or the condition of the ship or its equipment then further examination and testing may be conducted as may be necessary. In this respect reference is made to the IMO Survey Guidelines under the HSSC system, resolution A.1053(27) adopted on 30 November 2011 (see paragraph 22.3 of this manual)..

C   International Sewage Pollution Prevention Certificate, Certificate of Type Test for Sewage Treatment Plants

1   The inspector should examine the International Sewage Pollution Prevention Certificate and Certificate of Type Test for Sewage Treatment Plants to verify that they meet the requirements. As a preliminary check, it should be confirmed that the dates of surveys and inspections are still valid.

2   The inspector should then examine the sewage treatment plant itself to verify that it matches the description (serial number) on the certificates, and ensure that no unauthorized modifications or piping arrangements have been added.

3   If the State requires there be no overboard discharges, then the overboard discharge valve should be (have been) secured in an acceptable manner (locked closed).

4   The inspector should then verify that there are enough consumables, such as chlorine tablets, if used, on hand to ensure proper operation of the plant.

D   Garbage Record Book, Garbage Management Plan and Placards

In this section D, all references are related to the revised MARPOL Annex V, with an entry into force date of 1 January 2013. It should be kept in mind that residues of solid bulk cargoes are covered by the definition of garbage.

1   The Inspector should examine the Garbage Record Book and the Garbage Management Plan.

2   Ships of 400 gross tonnage and above, and ships certified to carry 15 persons or more, must have a Garbage Record Book on board.

3   The Garbage Record Book should be inspected to ensure that the records are up to date. The Book should show the date, time and position of the ship when garbage was discharged into the sea, discharged to reception facilities or incinerated. It should also contain the amount of garbage handled. The different types of garbage are divided into nine different categories, with plastic as category A. From 1 January 2013, the discharge of all garbage, including plastics, together with synthetic material such as ropes and fishing nets, and excluding food waste, is prohibited at sea, wherever the ship may be. Special attention should be paid to the handling of any type of garbage discharged at sea after 1 January 2013. When the Garbage Record Book states that garbage has been discharged to reception facilities, the master should provide the inspector with a receipt for the amount of garbage transferred.

4   The inspector should also check that all ships above 12 m in length display placards which notify crew and passengers of the legal discharge requirements for garbage.

5   Ships of 100 gross tonnage and above, and ships certified to carry 15 persons or more, must have a Garbage Management Plan on board. The plan should be inspected to ensure it contains procedures for collecting, sorting, processing and disposing of garbage, including identification of the person in charge of carrying out the plan.

6   For the residues of solid bulk cargoes it is important to identify whether or not the cargo residues are considered harmful to the marine environment. The seven parameters identified in paragraph 3.2 of the 2012 Guidelines for the implementation of MARPOL Annex V are important for making the correct assessment.

E   International Air Pollution Prevention Certificate and its Supplement

1   The inspector should examine the International Air Pollution Prevention (IAPP) Certificate including its Supplement and after 1 January 2013 also the International Energy Efficiency (IEE) Certificate should be inspected. As a preliminary check it should be confirmed that the required surveys have been performed.

2   It should be established from the Supplement to the IAPP Certificate which particular air pollution arrangements and equipment are installed on board and hence verify the availability and status of the following certification, record books, records and procedures, where applicable:

- Engine International Air Pollution Prevention Certificate including Supplement;
- Technical File and any approved amendments;
- Record Book of Engine Parameters;
- Documentation as required relating to other on-board NO_x Verification procedures;
- Approved Method File;
- Ozone-Depleting Substances Record Book;
- Fuel oil changeover procedures;
- Records of fuel oil changeovers;
- Documentation as required relating to approved equivalent means of complying with sulphur oxides and particulate matter emission controls;
- Approval certification in respect of vapour emission control system;
- Volatile Organic Compound Management Plan;
- Incinerator Type Approval Certificate;
- Incinerator operating Manual;
- Incinerator operator training records;
- Records of incinerator operating temperatures;

- Receipts from reception facilities in respect of ozone-depleting substances and equipment containing same or exhaust gas cleaning residues;

- Bunker delivery notes and associated representative samples records thereof;

- Notifications to the ship's flag State together with any available commercial documentation relevant to non-compliant bunker delivery; and

- Documentation relating to the approval of other alternatives or exemptions as provided for in Annex VI.

3   If the inspector, from general impressions or observations on board, has clear grounds for believing that the condition of the ship or its equipment do not correspond substantially with the particulars of the certificates or the documents a more detailed inspection should be conducted.

4   The more detailed inspection should cover both the air pollution arrangements or equipment which gave rise to the need to undertake that inspection and other on board air pollution prevention arrangements or equipment. In so doing it should also be established whether the master or crew are familiar with the various relevant procedures to prevent air pollution.

5   The more detailed inspection should also confirm that no unapproved modifications have been made to the air pollution arrangements or equipment or additional unapproved equipment has been installed.

6   Should any doubt arise as to the maintenance or the condition of the ship or its equipment related to air pollution control then further examination or testing may be conducted as necessary. In this respect reference is made to resolution MEPC.181(59), Revised guidelines for port State control under the revised MARPOL Annex VI.

Appendix 17
Investigations into contravention of discharge provisions

1   This appendix is intended to identify information which is often needed by a flag State to enable it to cause proceedings to be brought in respect of alleged violation of the discharge requirements, for the prosecution of such violations.

2   It is recommended that in preparing a port State report on deficiencies where contravention of the discharge requirements is involved, the authorities of the coastal or port State be guided by the itemized list of possible evidence as shown in the addendum to this appendix. It should be borne in mind in this connection that:

> **.1**   the report aims to provide the optimal collation of obtainable data; however, even if all the information cannot be provided, as much information as possible should be submitted; and

> **.2**   it is important for all the information included in the report to be supported by facts which, when considered as a whole, would lead to the justification for an alleged violation for the port or coastal State.

3   In addition to the port State report on deficiencies, a report should be completed by a port or coastal State on the basis of the itemized list of possible evidence. It is important that these reports are supplemented by documents such as:

> **.1**   a statement by the observer of the alleged violation. In addition to the information required under section 1 in appendix 18 of this manual, the statement should include considerations which lead the observer to conclude that none of any other possible pollution sources is in fact the source;

> **.2**   statements concerning the sampling procedures both of the slick and on board. These should include location of and time when samples were taken, identity of person(s) taking the sample and receipts identifying the persons having custody and receiving transfer of the samples;

> **.3**   reports of analyses of samples taken of the slick and on board; the reports should include the results of the analyses, a description of the method employed, reference to or copies of scientific documentation attesting to the accuracy and validity of the method employed and names of persons performing the analyses and their experience;

> **.4**   a statement by the inspector on board together with his rank and organization;

> **.5**   statements by persons being questioned;

> **.6**   statements by witnesses;

> **.7**   photographs of the discharge; and

> **.8**   copies of relevant pages of Oil Record Books, Cargo Record Book, Garbage Record Book, log-books, discharge recordings, etc.

4   The report referred to under paragraphs 2 and 3 above should be sent to the flag State. If the coastal State observing the pollution and the port State carrying out the investigation on board are not the same, the State carrying out the latter investigation should also send a copy of its findings to the State observing the pollution and requesting the investigation. If the alleged violation has taken place within the jurisdiction of the coastal State, this State may decide to cause proceedings by its own legal system.

5   All observations, photographs and documentation should be supported by a signed verification of their authenticity. All certifications, authentications or verifications shall be executed in accordance with the laws of the State which prepares them. All statements should be signed and dated by the person making the statement and, if possible, by a witness to the signing. The names of the persons signing statements should be printed in legible script above or below the signature.

Appendix 18
Itemized list of possible evidence of contravention of the MARPOL Annex I discharge provisions

1 Action on sighting oil pollution

1.1 Particulars of ship or ships suspected of a contravention

1.1.1 Name of the ship

1.1.2 Reason for suspecting the ship in detail

1.1.3 Date and time (UTC) of observation or identification

1.1.4 Position of the ship, and how this has been established

1.1.5 Flag and port of registry

1.1.6 Type (e.g., tanker, cargo vessel, passenger ship, fishing vessel), size (estimated tonnage) and other descriptive data (e.g., superstructure colour and funnel mark)

1.1.7 Draught condition (loaded or in ballast)

1.1.8 Approximate course and speed

1.1.9 Position of slick in relation to ship (e.g., astern, port, starboard)

1.1.10 Part of the ship from which the alleged illegal discharge was seen emanating

1.1.11 Whether the alleged illegal discharge ceased when the ship was observed or contacted by radio

1.2 Particulars of slick or other discharge

1.2.1 Date and time (UTC) of observation if different from 1.1.3

1.2.2 Position of the oil slick or other alleged illegal discharges in longitude and latitude if different from 1.1.4

1.2.3 Approximate distance from the nearest landmark in nautical miles

1.2.4 Approximate overall dimension of the oil slick or other alleged illegal discharge (length, width and percentage thereof covered by oil)

1.2.5 Physical description of the oil slick or other alleged illegal discharges (direction and form – e.g., continuous, in patches or in windrows)

1.2.6 Appearance of the oil slick (indicate categories) or colour of other alleged illegal discharges

Category A: Barely visible under most favourable light condition

Category B: Visible as silvery sheen on water surface

Category C: First trace of colour may be observed

Category D: Bright band of colour

Category E: Colours begin to turn dull

Category F: Colours are much darker

1.2.7   Sky conditions (bright sunshine, overcast, etc.), light fall and visibility (km) at the time of observation

1.2.8   Sea state

1.2.9   Direction and speed of surface wind

1.2.10   Direction and speed of current

1.2.11   Depth of water according to chart

1.3   Identification of the observer(s)

1.3.1   Name of the observer

1.3.2   Organization with which the observer is affiliated (if any)

1.3.3   Observer's status within the organization

1.3.4   Observation made from aircraft/ship/shore/otherwise

1.3.5   Name or identity of ship or aircraft from which the observation was made

1.3.6   Specific location of ship, aircraft, place on shore or other from which observation was made

1.3.7   Activity engaged in by observer when observation was made, for example, patrol, voyage flight *en route* from ... to ..., etc.

1.4   Method of observation and documentation

1.4.1   Visual

1.4.2   Photographs*

1.4.3   Remote sensing records and/or remote sensing photographs

1.4.4   Samples taken from the slick

1.4.5   Any other form of observation (specify)

1.5   Other information, if radio contact can be established

1.5.1   Master informed of alleged violation

1.5.2   Master's explanation (including what operations are being undertaken at that time)

1.5.3   Ship's last port of call

1.5.4   Ship's next port of call

1.5.5   Name of ship's master and owner

1.5.6   Ship's call sign and IMO number

---

* A photograph of the discharge should preferably be in colour. Photographs can provide the following information: that a material on the sea surface *is* oil, that the quantity of oil discharged does constitute a violation of the Convention, that the oil is being or has been discharged from a particular ship, the identity of the ship. Experience has shown that the aforementioned can be obtained with the following four photographs:

– details of the slick taken almost vertically down from an altitude of less than 300m with the sun behind the photographer;

– an overall view of the ship and slick showing oil emanating from a particular ship;

– a view which shows that there is no slick in front of the ship;

– details of the ship for the purpose of identification.

2   Investigation on board

2.1   Inspection of IOPP, CoF or NLS Certificate

2.1.1   Name of ship*

2.1.2   Distinctive number or letters and IMO number

2.1.3   Port of registry

2.1.4   Type of ship

2.1.5   Date and place of issue

2.1.6   Date and place of endorsement(s)

2.2   Inspection of supplement to IOPP Certificate

2.2.1   Applicable paragraphs of sections 2, 3, 4, 5 6 and 7 of the supplement (non-oil tankers)

2.2.2   Applicable paragraphs of sections 2, 3, 4, 5, 6, 7, 8, 9 and 10 of the supplement (oil tankers)

2.3   Inspection of Oil Record Book (ORB)

2.3.1   Copy sufficient pages of the ORB part I to cover a period of 30 days prior to the reported incident

2.3.2   Copy sufficient pages of the ORB part II (if on board) to cover a full loading/unloading/ballasting and tank cleaning cycle of the ship. Also copy the tank diagram.

2.4   Inspection of the NLS Certificate or CoF and the P & A Manual

2.4.1   List of products the ship is certified to carry

2.4.2   Limitations as to tanks in which these products may be carried

2.4.3   Ship equipped with an efficient stripping system

2.4.4   Residue quantities established at survey

2.5   Inspection of Cargo Record Book (CRB)

2.5.1   Copy sufficient pages of the CRB to cover a full loading/ unloading/ballasting and tank cleaning cycle of the ship. Also a copy of the tank diagram.

2.6   Inspection of the log-book

2.6.1   Last port, date of departure, draught forward and aft

2.6.2   Current port, date of arrival, draught forward and aft

2.6.3   Ship's position at or near the time the incident was reported

2.6.4   Spot check if positions mentioned in the log-book correspond with positions noted in the ORB or CRB

---

* If the ship is not issued with any certificate, as much as possible of the requested information should be given. If the ship does not have an IOPP Certificate, a description should be given of the pollution-prevention equipment and arrangements on board.

2.7   Inspection of other documentation on board

2.7.1   Other documentation relevant for evidence (if necessary make copies) such as:

–   recent ullage sheets

–   records of monitoring and control equipment

–   recent repairs to monitoring or control equipment

–   cargo documents of cargo currently or recently carried, together with relevant information of required unloading temperature, viscosity and/or melting point

–   records of temperature of substances during unloading

2.8   Inspection of ship

2.8.1   Ship's equipment in accordance with the supplement of the IOPP Certificate

2.8.2   Samples taken. State location on board.

2.8.3   Traces of oil in vicinity of overboard discharge outlets

2.8.4   Condition of engine-room and contents of bilges

2.8.5   Condition of oily-water separator, filtering equipment and alarm, stopping or monitoring arrangements

2.8.6   Contents of sludge and/or holding tanks

2.8.7   Sources of considerable leakage

2.8.8   On board of oil tankers the following additional evidence may be pertinent:

.1   oil on surface of segregated or dedicated clean ballast

.2   condition of pumproom bilges

.3   condition of COW system

.4   condition of IG system

.5   condition of monitoring and control system

2.9   Statements of persons concerned

2.9.1   If the ORB part I has not been properly completed, information on the following questions may be pertinent:

.1   was there a discharge (accidental or intentional) at the time indicated on the incident report?

.2   is the bilge discharge controlled automatically?

.3   if so, at what time was the system last put into operation and at what time was the system last put on manual mode?

.4   if not, what were the date and time of the last bilge discharge?

.5   what was the date of the last disposal of residue and how was disposal effected?

.6   is it usual to effect discharge of bilge water directly to the sea, or to store bilge water first in a collecting tank? Identify the collecting tank.

.7   have oil fuel tanks recently been used as ballast tanks?

2.9.2   If the ORB part II has not been properly completed, information on the following questions may be pertinent:

.1   what was the cargo/ballast distribution in the ship on departure from the last port?

.2   what was the cargo/ballast distribution in the ship on arrival in the current port?

.3   when and where was the last loading effected?

.4   when and where was the last unloading effected?

.5   when and where was the last discharge of dirty ballast?

.6   when and where was the last cleaning of cargo tanks?

.7   when and where was the last COW operation and which tanks were washed?

.8   when and where was the last decanting of slop tanks?

.9   what is the ullage in the slop tank(s) and the corresponding height of the interface?

.10   which tanks contained the dirty ballast during the ballast voyage (if ship arrived in ballast)?

.11   which tanks contained the clean ballast during the ballast voyage (if ship arrived in ballast)?

2.9.3   In addition, the following information may be pertinent:

.1   details of the present voyage of the ship (previous ports, next ports, trade)

.2   content of oil fuel and ballast tanks

.3   previous and next bunkering, type of oil fuel

.4   availability or non-availability of reception facilities for oil during the present voyage

.5   internal transfer of oil fuel during the present voyage

2.9.4   In the case of oil tankers, the following additional information may be pertinent:

.1   the trade the ship is engaged in, such as short/long distance, crude or product or alternating crude/product, lightening service, oil/dry bulk

.2   which tanks are clean and which are dirty

.3   repairs carried out or are envisaged in cargo tanks

2.9.5   Miscellaneous information:

.1   comments in respect of the condition of ship's equipment

.2   comments in respect of the pollution report

.3   other comments

3   Investigation ashore

3.1   Analyses of oil samples

3.1.1   Indicate method and results of the analyses of the samples

3.2   Further information

3.2.1   Additional information on the ship, obtained from oil terminal staff, tank cleaning contractors or reception facilities may be pertinent

4   Information not covered by the foregoing

5   Conclusion

5.1.1   Summary of the investigator's conclusions

Any information under this heading is, if practicable, to be corroborated by documentation such as signed statements, invoices, receipts, etc.

5.1.2   Indication of applicable provisions of Annex I of MARPOL which the ship is suspected of having contravened

5.1.3   Did the results of the investigation warrant the filing of a deficiency report?

Appendix 19

Guidance in respect of the inspection of marine diesel engines in order to determine compliance with the requirements of regulation 13 of Annex VI

Further to the guidance given in paragraph 21.7 of this manual, the following notes provide some further explanation of the approach which may be adopted by inspectors in order to determine whether a marine diesel engine is in a duly compliant condition with the requirements of regulation 13.

As given in chapter 14, each engine which has been certified in accordance with the requirements of the NO_x Technical Code 2008, or the earlier 1997 version of the same, it is required to have an approved Technical File on board. This Technical File is approved as part of the process leading to the issue of an Engine International Air Pollution Prevention (EIAPP) Certificate, normally prior to the first installation of the engine on board of a ship. This EIAPP Certificate is valid for the service life of the engine while installed on a ship under the authority of the issuing flag State. Consequently, in view of the technical or other developments which may take place over that time it may be necessary that amendments to that Technical File have to be made and approved, such amendments must be considered as part of, although separate to, the Technical File as initially approved and also be on board and available as required.

The Technical File is the key document in assisting both the ship's crew to retain that engine in a compliant condition and the surveyor or inspector in verifying that compliant condition indeed is the case. If there is no Technical File on board – the survey or inspection is judged and terminated as uncompleted and due action ensues.

Where a more detailed inspection of an engine is to be undertaken, see appendix 15 section E, the procedure followed may typically correspond to that which would be applied by a surveyor at the time of a renewal, annual or intermediate survey. To assist in visualizing that process a flowchart based on figure 3 of appendix II of the NO_x Technical Code 2008 is reproduced in figure 12 (note, this does not cover the issues related to whether an Approved Method has been applied as may be required by regulation 13.7 of Annex VI).

As a general point it would be highlighted that it is for the ship's crew to show that an engine is compliant – as prompted by the inspector's questions – rather than for the inspector to determine that the engine is compliant. In the first instance the inspector should request the identification, purpose and location of all engines on board; for a ship with International Air Pollution Prevention (IAPP) Certificate this information is given in section 2.2.1of the Supplement. While on board the inspector should of course be vigilant as to other, unlisted or unadvised, engines.

In following this procedure for a particular engine the first four decision diamonds of the flowchart cover whether the engine actually requires to be certified. In this it is important to include in the assessment any engines installed outside the engine room itself, particularly those installed after the ship's construction. "Not installed" engines would be those typically wheel or trolley mounted and used for some aspect of ship operation or maintenance, for example engine driven shot-blast compressors. With regard to the "emergency use only" engines; this applies only to engines used to counter emergency situations on the ship onto which they are installed – regulation 13.1.2.1. In the case of engines which power some aspect of sea-bed mineral activities it is necessary that the engines do not also supply power for other purposes (i.e., propulsion, electrical power to accommodation or for general engine room purposes, hotel services, etc.), if so certification is required. In the case of ships constructed before 1 January 2000 it is necessary to verify that engines have not been subject to "Major Conversion", a point which, being retrospective, may have been overlooked by those arranging or undertaking such work. In this respect examination of recent engine work as contained in the machinery records would be an appropriate starting point.

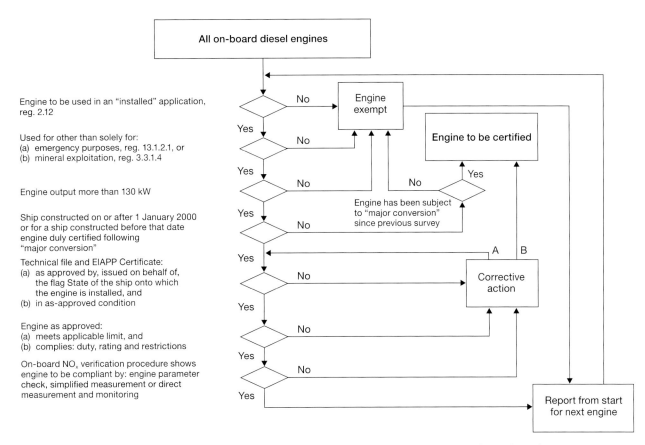

Figure 12 – *Flowchart based on figure 3 of appendix II of the NO$_x$ Technical Code 2008*

The fifth decision diamond covers the engine's EIAPP Certificate and Technical File. Both should be approved by, or on behalf of, the appropriate flag State. It should be noted that virtually all engine certification work is delegated to Recognized Organizations (see chapter 23). This documentation must refer to the engine design / type and basic arrangement details together with the engine's serial number as found on the nameplate fixed to the engine.

The Technical File is an approved document and should therefore not be subject to unauthorized amendments, deletions or additions. Despite being used on a regular basis during normal operations of the ship, the document must be retained in a good condition with any applied seals kept intact. Inspectors should therefore generally accept offers from the ship's crew to photocopy relevant sections of the Technical File which can then be used in the engine room, or elsewhere, thereby protecting the integrity, and cleanliness, of this official document.

The sixth decision diamond addresses whether the engine meets the required limit and whether it has been installed in accordance with its approved duty, rating and restrictions. In respect of the first part of this question this should normally not be a point to be verified as part of an inspection since this aspect should have been correctly addressed as part of the flag State surveys. However for completeness, it should be mentioned that the applicable limit value is given in 1.9.5 of the Supplement to the EIAPP Certificate (Note, the arrangement of Supplement to EIAPP Certificates issued in accordance with the 1997 version of NO$_x$ Technical Code is different to those issued under NO$_x$ Technical Code 2008). The engine's certified emission value is given in 1.9.6 of the Supplement to the EIAPP Certificate, where an engine has been certified to more than one duty cycle, see below, the certified emission value for each duty cycle will be given.

In respect of duty there are four possible test cycles for which an engine can be certified to in accordance with the NO$_x$ Technical Code 2008. These are given in appendix II to Annex VI and in chapter 3 of the NO$_x$ Technical Code 2008 and are:

E2   Main (propulsion) engines operating at constant speed. This also includes engines installed in diesel-electric type drive arrangements and those which drive controllable pitch propellers. In the latter category are engines with combinator type controls whereby both engine speed and propeller pitch can be altered within the bounds of the permitted torque limits and therefore this category is, potentially confusingly, not confined only to constant speed engines.

E3   Main or auxiliary engines – propeller-law-operated. These are engines typically driving fixed-pitch propellers, or pump types, where the speed/torque characteristic is fixed.

D2   Auxiliary engines operating at constant speed. These are typically the ship's generator engines although there can be other applications with constant speed drive requirements.

C2   Auxiliary engines operating a variable speed and load. This cycle is not commonly used, typically it applies to engines driving certain large pumps as may be installed as part of dredging systems.

As above, some engines with duplicate applications are certified to more than one test cycle. The installation of an engine in accordance with the relevant duty cycle or cycles need to be verified at the ship's initial survey or, in the case of retrofits, subsequently when the engine was installed.

Therefore the question should be posed whether the main engine is, or engines are, duly certified to the appropriate E2 or E3 cycle? And whether the engine driving the generator currently supplying the ship's electrical power is certified to the D2 cycle?

Rating covers both the rated power and the rated speed of an engine. The values as given on the EIAPP Certificate and in the Technical File should therefore correspond to those as given on the engine's nameplate and can be checked against the engine's actual performance or logbook records. For example, is a D2 cycle engine certified for a rated speed of 720 rpm being used to drive a 900 rpm alternator? Some engine certification can be for a range of powers (kW/cylinder) or rated speeds, however, the engine's actual value must be within the given range.

Applicable restrictions will be given in the engine's Technical File. Commonly this refers to the charge (scavenge) air cooling arrangements – the higher the air temperature the greater the tendency to NO_x formation. Hence an engine may be approved for direct sea-water cooling of the charge (scavenge) air or have fresh water cooling – in the latter case for any given condition the temperatures will be higher. Additionally there are air cooled/radiator engines. With such aspects the inspector may take the questioning line of "show me how the restrictions are complied with".

The seventh question point refers to the application of the on-board NO_x verification procedure. Of the three options available, 2.4.3 of the NO_x Technical Code 2008, the parameter check method is applied to all engines as constructed and, to date, has been generally retained as the method by which through life compliance is demonstrated in practice and it is therefore this method that is the basis of these guidance notes.

While there exists a wide variation in the format and arrangement of Technical Files, each is required to contain the information as listed in 2.4.1 of the NO_x Technical Code 2008 and under sub-point .4 of that listing is the on-board NO_x verification procedure which will provide all necessary details relating to the particular parameter check method to be applied. Under the parameter check method given for a particular engine there will be listed the NO_x critical components and the NO_x critical settings and operating values.

The starting point for the verification of the Parameter Check method will always be the mandated Record Book of Engine Parameters, 6.2.2.7.1 and 6.2.2.8 of the NO_x Technical Code 2008. There is no prescribed format for this record, however, engine builders will often have provided example pages within the Technical File which the ship's crew should have copied as required; the pages in the Technical File itself should not have been completed. As mentioned above, this is a statutory document which must not be subject to unauthorized additions or other amendments. In some instances the engine builder will have provided a computer spreadsheet file which is to be duly completed at each occasion and from which pages can then be printed for retention and review at surveys or inspections.

As given in the referenced sections of the NO_x Technical Code 2008, the Record Book of Engine Parameters is to be completed for all actions taken relative to an engine's NO_x critical components, settings and operating values. This includes all associated like-for-like replacements and adjustments. Clearly a readily available, well presented Record Book of Engine Parameters will be a positive sign of an engine retained in a compliant condition.

The NO_x critical components typically divide into three categories: fuel injection system, charge (scavenge) air system and combustion chamber. Within each of these categories the engine certifier will have required certain components to be listed, these will differ between engine types and designs but commonly cover:

Fuel injection system
 Fuel injection nozzle
 Fuel pump
 Fuel cam

Charge (scavenge) air system
 Turbocharger type /model /number
  Compressor wheel
  Diffuser
  Turbine rotor
  Nozzle ring
 Cooler
 Combustion chamber
 Piston
 Cylinder cover / head
 Liner (not generally for trunk piston engines)
 Piston rod (crosshead engines)
 Connecting rod (trunk piston engines)
 Aspects affecting compression pressure including valve timing

For engines with more complicated operating arrangements there will be other relevant components listed.

The Technical File will list the required marking or other defining characteristic (such as the uncompressed thickness of cylinder head sealing rings) which apply to the approved NO_x critical components. In some instances there may be more than one option: pistons crowns for crosshead engines, for example, may be manufactured in a number of different ways, each with its own marking; however, in NO_x compliance terms all are equivalent. Usually the Technical File will also include diagrams as to where these marking are actually located on the respective components. The ship's crew should be able to demonstrate that the fitted components are duly marked, or other defining characteristics are, as required.

Reference to recent entries in the Record Book of Engine Parameters would be a typical starting point in this verification. In the case of auxiliary generator engines the inspector may confine the verification to an idle engine taking that to be representative of the others if all indications are that the engines are identical. However, should non-compliance be determined, the inspection would be extended to cover all engines even if this means load being transferred off a running engine which then needs to be shut down. Alternatively, the markings on spare parts may be checked   these of course need to be truly representative of those intended to be used, not simply an exhibition set maintained for survey/inspection purposes. Engine builders have generally gone to some trouble to ensure that these markings are normally readily accessible (albeit that some covers or crankcase doors may need to be removed/opened). Where components are not readily accessible, for example those on turbocharger internal components or on nested camshafts of Vee configuration engines, the inspector's judgment should be applied as to whether such components could have been changed since the engine was constructed or the last extensive survey and hence need to be verified. Indeed an inspection of the engine's maintenance records would be key to assessing the likelihood that engine components have been changed for non-compliant options. Of all the components to be verified, and generally the simplest to verify, is the fuel injection nozzle which is also usually the most crucial to NO_x emissions.

The most common reasons for non-compliance will be the fitting of components which are either not duly marked or not marked as required. The engine builder will normally have completed the Technical File in terms of the markings they, or the licensor, applies; components obtained from other sources will not be so marked. While such other components are not unacceptable in principle, provided they do not adversely affect NO_x emissions, the shipowner, or other party wishing to use same, will have needed to have demonstrated that these do not compromise the engine's NO_x emission characteristics and to have had the relevant defining characteristics or markings approved, typically as approved addendum pages as part of, but separate to, the Technical File as initially approved.

The Technical File will also specify the NO_x critical settings and operating values and whether the given values are maximums or minimums. Again these differ between engine types. In terms of fuel pump timing, the crucial element in terms of NO_x compliance, many trunk piston engines will be limited in terms of the static timing before top dead centre (or fuel pump lift at top dead centre). In these cases a maximum advance limit (or lift) value will be given and hence any value below that limit value would be acceptable; the less the timing is advanced the lower the resulting NO_x emissions. Conversely, fuel pump timing for crosshead engines may be limited in terms of maximum combustion pressures across the load range and it is this which is the defining characteristic; the settings are whatever values ensure that this pressure line is not exceeded. Engines with electronic control of fuel pump timing will typically have defined and identifiable data sets which are specified in the Technical File.

Other NO_x critical settings and operating values would typically be related to aspects such as charge (scavenge) air temperatures (where due allowance needs to be made for load and, depending on system, ambient conditions), compression pressure (although this will be also partly controlled by the NO_x critical components) and back pressure. Other settings or operating values relevant to the particular design will have been included as part of the development of the Technical File leading to its approval. Typically the Technical File will outline the procedures by which all these NO_x critical settings or operating values are to be demonstrated.

Again the Record Book of Engine Parameters would normally be the starting point of any survey or inspection verification of NO_x critical settings or operating values. As with the components, in assessing which need to be verified the inspector should be guided by the likelihood of change and the work which has recently been undertaken on the engine. In the case of the operating values, such as main engine combustion pressures, these of course require the engine to be operating which will not normally be the case during an inspection which typically takes place while a ship is berthed. In this the inspector's judgment is required as to whether past records, as included in the Record Book of Engine Parameters, are a reliable indicator of the engine's actual performance taking into account other information such as that recorded in the engine room log book or engine management systems. It of course being no small point to require a ship to put to sea in order to work the main engine up to the 50–75% load range where these combustion pressures could be checked! Fuel pump lift values and exhaust valve timing can instead be checked with the engine idle against those given for the relevant operating values in the latest entry to the Record Book of Engine Parameters.

From the fifth, sixth and seventh question points there is either a compliant (Yes) route or a non-compliant (No) route. The latter all lead to the corrective action box. In some instances, it will be possible for the ship's crew (or service personnel) to make (with varying degrees of effort and cost) the required changes in order to bring the engine(s) back into compliance – the "A" route out. However, in other cases – the "B" route out – this will not be possible and in such extreme cases it may require new certification of the engine(s).

The above represents a thumbnail sketch of the inspection procedure as may be applied to an engine. Over course of time there will be developing experience together with changes of engine designs and arrangements which will of course influence the actual procedure – however in all cases it will be driven by the procedures and arrangements as given in the approved Technical File.

Notes

Notes

Notes

Notes

Notes

Notes

Notes

Notes

Notes

Notes